# The Chicken Coop Manual

*By Kathryn Kerby*

ISBN-13: 978-1732469709

*Published by*
Farm and Ranch Success Publications
*a branch of*
Frog Chorus Farm
Snohomish, WA
www.frogchorusfarm.com

*With help from*
Ingram Spark Print on Demand Services
www.ingramspark.com

*And*
Zamzar File Conversion Services
www.Zamzar.com

I regard it as just as desirable to build a chicken house as to build a cathedral.
- Frank Lloyd Wright



Chickens are the new dog.
- SR Rubino



The cock may crow but it's the hen who lays an egg.
-Margaret Thatcher



Back when Herbert Hoover promised Americans a chicken in every pot and a car in every garage, he could not have guessed that what people really wanted was a chicken in every garage.
-Michael Jamison

# *ACKNOWLEDGEMENTS*

This guide would have been impossible without the very generous support and guidance of so many people. First to my family and my husband, who always encouraged me to go conquer that next big unknown. To all my mentors, particularly my holistic veterinarian Dr. Douglas R. Yearout of *All Animal and Bird Clinic*, Lake Stevens, Washington, for teaching me how to keep birds healthy, what's normal in birds and what needs attention. To all the staff members of the Washington State University/Snohomish County Extension office and Snohomish County Conservation District, for answering 1001 questions over the years about livestock management, animal waste and compost handling systems, water pollution control and runoff mitigation, and animal housing designs which won't collapse during the next windstorm, ice storm or snowfall. A big thanks to the Snohomish Health District for ideas on rodent control and disease prevention and for working with us when our rat issue got away from us. A big thank you to all the folks who agreed to allow their photos to be used in this manual; those credits are listed separately. A very big thank you to the Louisiana State University's AgCenter for not only maintaining such an excellent library of animal housing plans, in particular the poultry housing collection, but also for allowing me to use those plans as examples in this book. And finally, a big fond thank you to all the birds which have graced our farm with their beauty, their production, their antics, their personalities and their voices over the years. The farm would be an empty, sterile, hollow place without them.

# *DISCLAIMER*

This guide is intended to provide ideas for how to build a chicken coop which will provide adequate service to a small flock owner. It is not an engineering manual, and the designs, approaches, methods and plans discussed here are described and intended as suggestions only. The author makes no guarantee or warrantee of any kind, that any specific piece of information will be applicable to any particular reader's unique circumstances. Furthermore, the author strongly encourages every reader to verify that any design, plan or method is in full compliance with local, state, provincial and/or federal laws as may be relevant. Finally, the author strongly believes that readers must do their own homework to ensure that their planned poultry structures will serve their needs. In other words, don't be an idiot and do something stupid and then try to blame the book for that error in judgment. When in doubt, consult with local experts and authorities. The relevant county Extension office, Conservation District, NRCS office and/or the agricultural department of the nearest land grant university are excellent resources for what works locally. Google works pretty darn well too.

# *COPYRIGHT INFORMATION*

Copyright © 2014, Kathryn Kerby
All rights reserved. Without limiting rights under the copyright reserved above, no part of this publication may be reproduced, stored, introduced into a retrieval system, distributed or transmitted in any form or by any means, including without limitation photocopying, recording, or other electronic or mechanical methods, without the prior written permission of the publisher, except in the case of brief quotations embodied in critical reviews and certain other noncommercial uses permitted by copyright law. The scanning, uploading, and/or distribution of this document via the internet or via any other means without the permission of the publisher is illegal and is punishable by law. Please purchase only authorized editions and do not participate in or encourage electronic piracy of copyrightable materials. For permission requests, contact the author at www.frogchorusfarm.com.

# TABLE OF CONTENTS

| | |
|---|---|
| ACKNOWLEDGEMENTS | 3 |
| DISCLAIMER | 3 |
| COPYRIGHT INFORMATION | 3 |

## INTRODUCTION — 6
| | |
|---|---|
| Building The Perfect Chicken Coop | 6 |
| Purpose, Appearance, Efficiency | 7 |
| Regulations, and the Renegade Chicken | 7 |
| Book Section Descriptions: | 8 |

## SECTION I: COOP DESIGN CONSIDERATIONS — 10
| | |
|---|---|
| A Look At Coop Design Criteria | 10 |
| Coop Size | 12 |
| Coop Furniture: Feeders & Waterers | 13 |
| Coop Furniture: Roosting Areas | 17 |
| Coop Furniture: Nest Boxes and Egg Laying | 19 |
| Coop Furniture: Brooders and brooding | 24 |
| The Human Factor – | 25 |
| Working Conditions, Efficiency, Sanitation | 25 |
| Protection From Weather | 29 |
| Vermin Control | 30 |
| Predator Control | 35 |
| Disease Control | 45 |
| Heating and Lighting | 55 |
| Bedding and Waste | 62 |
| Fixed or Mobile? | 66 |
| Chicken Tractors | 73 |
| Pre-Made Chicken Coops | 75 |
| Durability | 76 |
| Conclusion | 77 |

## SECTION II: CONVENTIONAL COOP PLANS — 78
| | |
|---|---|
| Non-Walk-in Coops | 79 |
| Walk-in Coops | 80 |
| Plan #1: Non-Walk-In Coop For 8-16 Birds | 83 |
| Plan #2: 20 to 30 Bird Family Flock Coop | 85 |
| Plan #3: 25 to 36 Bird Coop | 87 |
| Plan #4: 25 to 40 Bird Coop | 89 |
| Plan #5: 40 to 50 Bird Small Farm Flock | 91 |
| Plan #6: 50 to 80 Bird Small Farm Coop | 93 |
| Plan #7: 100 Birds Farm Flock | 95 |
| Plan #8: Field Shelter for 100 Bird Farm Flock | 97 |

## SECTION III: ALTERNATIVE COOP DESIGNS — 99
- Car Canopy — 101
- Greenhouse/Hoop House — 104
- Truck Bed With Canopy — 107
- Shipping Crates and Containers — 110
- Horse or Livestock Trailer — 112
- Cattle Panels and Tarps — 114
- Metal Erector Frames With Tarps — 117
- Pallets — 121
- Rebar Frame — 124
- Chain Link Dog Kennels — 127
- Geodesic Domes — 130
- Converted Barn Stall — 134

## SECTION IV: SUPPLEMENTAL INFORMATION — 136
- A Few Last Comments — 136
- Additional Resources — 137
- Conclusion — 142
- ABOUT THE AUTHOR — 143
- PHOTO AND ILLUSTRATION CREDITS — 143

# INTRODUCTION

*One of the author's home-bred hens, on a path through the woods. I surprised her mid-stride.*

## *Building The Perfect Chicken Coop*

Say the phrase "chicken coop" and almost everyone will be able to picture a coop. A home for chickens, within which they eat, drink, sleep and lay eggs. While a great majority of modern society has very little working memory of life on a typical diversified farm, somehow this one tidbit of traditional knowledge has survived. If an individual wakes up one day and decides he or she needs a chicken coop, most of the rest of the world will know what that is.

That situation is happening all over the country. The causes and motivations for folks wanting their own poultry flock are beyond the scope of this publication, but suffice to say that both urban-type backyard flocks, and small working commercial flocks, are both increasing in numbers. According to USDA's Census of Agriculture information, commercial flock production rose between 2002 and 2007 (the latest data available from that source). In 2002, poultry and egg sales combined rose by 55%. While impressive, the real surprise has been the explosive growth of urban chicken flocks. A variety of urban centers and suburban metropolitan population centers have seen a resurgence in residents wanting small flocks for their own use.

One of the first questions which wannabe poultry owners must ask, and answer, is simply: what sort of coop do I want and need for my new flock? This publication will try to provide at least some of the many potential answers.

## *Purpose, Appearance, Efficiency*

There are a lot of different ways to build a perfectly serviceable chicken coop. The size, type and layout will be determined by things like the size of the flock, the weather conditions throughout the year in the given location, predator issues, and budget issues. Some wannabe poultry owners will also either want, or need to follow, certain guidelines for appearance, thanks to neighborhood covenants or regulations about how outbuildings need to look. And finally, no matter what the above criteria are, a chicken coop should be efficient, in terms of materials acquisition, construction, usage and even re-usage. When I say "efficient", I mean that the design should provide for both the birds' needs as well as the owner's needs, throughout the year, in an affordable, durable way. There are many different ways to accomplish these goals, and a lot of ways to re-purpose materials a prospective flock own might already have, or can acquire inexpensively. We'll talk about all these factors in the paragraphs to come.

## *Regulations, and the Renegade Chicken*

One issue which all wannabe chicken owners should address, is the legality of owning chickens AND putting up a coop in their particular area. Both of those topics vary widely, from a complete lack of regulatory oversight, to needing permits, to being specifically illegal. To make things even more interesting, regulations can vary dramatically from one subdivision or neighborhood to the next, and one town to the next. While multiple municipal governments are debating whether to allow chickens, some merely prescribe how many, or whether roosters are allowed, or how much space each bird should have. Other locations get into much more specific guidelines about the types of coops which are permitted, the size, the design, and the features within (for instance, some will forbid, or require, certain features). Some subdivisions will go to the length of saying that the coop's architectural style, detailing and color must match the primary residence. Whatever the local requirements, wannabe chicken owners should become familiar with those requirements, so that there is no question about what is or is not permitted. Once those details are known, coop design features can be selected without worrying if the local authorities are going to object. We'll consider options for how to meet those criteria, whatever they may be.

On the other hand, many people are engaging in renegade poultry keeping, for a variety of reasons. They may not know or care about local regulations. They may know the regulations very well, and are actively trying to change them. They may simply want their own flock with a minimum of visible evidence, and keep a very low profile. Chickens are one of the few common livestock species which are small enough to be kept in apartments, but yes some considerations must be given to ultra-small living situations. As a formal statement of intent, I don't want to encourage folks to break the law, whatever the law might be in their area. I will discuss how to manage poultry as cleanly, quietly and unobtrusively as possible, because that is desirable regardless of whether a flock is legal. It's up to the reader to decide whether to follow or ignore the applicable regulations.

## *Book Section Descriptions:*

This guide is set up in three different sections, so that folks with different goals, challenges or amounts of experience, can jump right to the information that most closely meets their needs:

Section I is mostly narrative, where I talk about design features and why they're important. This section is for folks who want to get started with birds but have no experience with poultry. We'll talk about why roosts are important, why nest boxes are set up the way they are, why feeders and waterers are set up they way they are, etc. However, if the reader prefers to just see the drawings without a lot of chatter, feel free to turn to either Section II or Section III later in this guide to get right to the details.

Section II is a series of conventional stud construction coop plans, from a well-respected American agricultural university. More details on that university's plans, are available in that section. These plans are typically free, in PDF format, available for immediate download. These plans have been used for decades either as direct plans, or as the basis for plans, for both hobbyist and commercial flocks. While these plans do not have professional engineering stamps on them (which some building/planning departments now require) they could be used to generate such plans if needed.

Section III is a series of alternative coop designs, showing how common materials can be scrounged or re-purposed for service as a chicken coop. This section is only an introduction to the possibilities; there are countless others. But a survey of these alternatives will provide a wide range of functional, free or inexpensive, often recycled materials which would have been thrown out or junked, yet are now serving as perfectly acceptable chicken coops. While the appearance of these coops might be a little unconventional, they work, they were inexpensive or free for the making, and sometimes they provided shelter where no other shelter options existed. I include them here to get the reader's creative juices going and to prove that if a person really wants a chicken coop, there are a lot of ways to create a good one.

Section IV is where we wrap up with some concluding comments. I'll also provide some follow-up websites and supplemental information.

It is my sincere hope that readers of this manual will find what they need, within these various sections.

# SECTION I: COOP DESIGN CONSIDERATIONS

*One of the author's home-grown hens raising a brood of her own in our hay shed.*

## *A Look At Coop Design Criteria*

Chickens have been kept for many generations, in many types of housing, all over the world. They've been kept in a bewildering array of coops, hutches, cages, enclosures, and other containers. Even within modern times these adaptable birds have survived, and thrived, in a wide array of housing. In other words, there are many ways to house chickens well. As long as their housing provides some basic needs, the birds will do well. So let's take a look at those needs, and how to meet them.

Chicken coops must provide some fairly universal functions, regardless of where they're located and how big the flock might be:

- Coop Size – depends on flock numbers and breed
- Furniture – roosts, feeders and waterers, nests and egg-laying, and brooders
- Human factor – working conditions, efficiency, sanitation
- Protection from weather
- Vermin control
- Disease control
- Predator control
- Heating and lighting
- Egg laying
- Bedding and waste management
- Fixed or mobile?
- Chicken Tractors
- Pre-Made Coops
- Durability

The good news is, flock owners have a lot of different ways to meet those needs. The other good news is, chickens are fairly accommodating and can succeed in a wide variety of settings. So let's take a closer look at each of those needs, what some of the options might be, and how those options compare.

For the sake of simplicity, from this point forward I'll make some assumptions:

- We will restrict our conversation to chickens of any breed. Banties, standards, and giant chicken breeds can all be housed in the following coop designs. Non-chicken fowl such as ducks, geese, pheasants, guineas and turkeys will not specifically be addressed. However, many of the designs we'll talk about for chickens, could be (and have been) successfully used for those species as well. I would encourage folks interested in those species, to also read up on their particular requirements and design the coop(s) accordingly.

- Most of these birds are adults, thus don't need supplemental heat. I will mention brooders and brooding chicks in passing in a few different locations throughout the text. However, the focus is on birds old enough to not need that supplemental heat.

- These birds are layers (as opposed to being primarily meat birds), such that we need nest boxes. Most of these designs could also be used for broiler production, and a few are ideal for that purpose. In that case, simply eliminate the nest boxes.

- These birds will have access to an outside yard of some kind.

Given those assumptions, we can start to figure out some "rules for the road" for our coop designs.

## *Coop Size*

For folks considering their first chicken coop, the first decision to make is: how many birds? Sometimes this number is determined by the number of eggs a household wants to have each day. The rule of thumb for laying hens is two eggs every three days, or approximately 4 eggs per week, per bird. If a family goes through a dozen eggs per week, that would be 12 eggs per week, or roughly 2 eggs per day (to allow for occasional breakage). To get those dozen eggs per week, a family would need 3 birds. One thing to keep in mind is that once a family has eggs every day, consumption of eggs often goes up. So where a family might have habitually purchased one dozen per week for years, seeing those beautiful fresh eggs in the nest box may encourage higher consumption rates. On the other hand, folks should allow for some buffer. Chickens are living creatures and they have off days. They get sick, they get injured, they get cold or flustered or 101 other things that can knock them off their egg laying schedule. So if a family absolutely positively wants a certain number of eggs per week, get slightly more birds than would minimally be required to provide that number.

Sometimes a family simply wants a small flock of a certain size, because they enjoy the birds and want some variety. They may want a pair of this and a pair of that, or a trio of something else. In that case, the number of eggs per day or per week isn't as important as having that variety. Sometimes families get into poultry due to their participation in 4H, FFA or other youth group activities. For instance, many young members of 4H want a trio of two hens and a rooster so they can show at fair that summer, and the rarer the breed, the better. In that case, the number of birds they get may be defined, recommended or otherwise suggested by that group.

Unfortunately, sometimes regulations will decide how many birds a family can have. Many towns and cities are changing their rules and allowing birds back within city limits. While that is generally good news, those rules often put a population cap on how many birds any given household can have. The numbers vary widely – 3 birds, 4 birds, 6 birds, fewer than 10 birds, etc. Even rural areas will sometimes have limits, albeit larger limits, for poultry populations on any given farm or residential property. It's tempting to ask neighbors how many birds are allowed, because it's more comfortable talking to neighbors than to a bureaucrat. However, neighbors rarely have up to date information on that unless they happen to work in the Planning Department. Happily, many cities, towns and counties have their regulations posted online, so it becomes a matter of searching those regulations to find the applicable rules.

Once the number of birds has been determined, the next considerations are the allowances for four items: roost space, nest boxes, feeders and waterers. Flock size will define how many nest boxes, how many waterers, and how much feeder space/roost space a coop must provide for comfortable housing. So let's take a moment to talk about "chicken furniture".

## *Coop Furniture: Feeders & Waterers*
Chickens are omnivores, which mean they can and will potentially eat a lot of different things. And chickens have been fed a huge variety of items as a regular part of their diet. That variety is entirely dependent upon their location, the ease with which certain foods or feeds can be acquired, the availability of edible items within the flock's reach, and the goals of the flock owner. For instance, many flocks are free-ranged in small yards or paddocks where they have the opportunity to eat whatever greenery might be there, as well as any bugs or small animals they can catch. Some new flock owners are surprised to see that chickens will exhibit hunting behaviors. Poultry can and will chase down and injure or kill things like

mice, rats and small snakes. Not reliably enough to be considered pest control for these species, but enough that a flock owner will occasionally find a dead something in their yard. Unfortunately, poultry don't draw the line there. They will also pester, bully, injure and kill young puppies, kittens, ducklings, goslings, and even chicks. Put all these factors together, and a chicken coop (and/or poultry yard) must be able to feed the birds, without unintended casualties along the way.

When setting up a chicken coop for the first time, wannabe poultry owners will need to consider what they want their birds to eat. Many folks intend to buy in whatever feed the birds need. This feed generally comes in three forms: small cylindrical pellets, where each pellet is a homogenous blend of foods, small crumbles, which are also a homogenous blend of foods but in an easier-to-eat shape, and scratch. Scratch is generally whole or cracked grains, and typically is composed of wheat, corn, and/or barley. That type of feed gets the name scratch because chickens will rake through it with their feet, i.e. "scratch it around", and peck out the bits that appeal to them most.

This habit of scratching or raking through feed with their feet, is a hallmark behavior for poultry and any feeder design will need to either allow for it, or be designed to minimize it. If owners are feeding their birds bought-in pellets, crumbles or scratch, the feeders will need to keep that feed off the ground so it's not wasted, keep it dry so it doesn't spoil, sprout, dissolve or ferment, and keep it away from rodents who definitely enjoy a free meal of bird feed. Two very common types of feeders are a bin feeder and tray feeder. The hanging feeder may be either round or rectangular, sized anywhere from holding a few dry quarts of material, to several gallons. It has a central bin or cylinder filled with feed, then a tray at the bottom which the birds eat out of. If this tray is mounted anywhere from 3" – 12" above the ground, birds can reach it with their beaks but will tend not to scratch the feed out with their feet. That results in a nice, clean feeding area. Depending on the size of the bin, this also allows for feeding once every few days. That latter point, however, requires that the feed be kept absolutely dry, and thus many locations will require that the feeder actually be inside the coop or otherwise protected from the elements. The other type of feeder, the tray feeder, is also typically kept off the ground, either by having its own stand, or by being mounted on a wall or upright stand of some kind. Here again, the tray portion will be 3" – 12" above ground, so that the birds can reach it but can't rake the feed out with their feet. The main difference between the bin feeder and the tray feeder, is that the bin has greater capacity. The advantage to that is that it doesn't need to be filled as often. The disadvantages are first that the contents must be absolutely dry, and secondly that rodents can easily get into the bin and eat the bird feed. If the poultry owner is feeding a wet feed, or has a known rodent problem, the tray feeders can work better. The feed is simply poured into the tray in measured amounts so that the birds only get as much as they can reasonably eat within about a 30 minute time span. This prevents any feed from spoiling, and it ensures that any opportunistic rodents don't get to dine out on the poultry owner's nickel.

In addition to the above types of feed, many folks also feed things like weeds from the garden, vegetable trimmings from the kitchen, stale bread, meat scraps, and even offal from other livestock butchering activities. All of these items are very much within the realm of reasonable and safe feeds for chickens. However, certain principles should be followed to ensure that these food types don't become a health hazard, eyesore or rodent magnet. For things like weeds from the garden and vegetable trimmings or stale bread from the house, feeding in the yard is often cleaner than feeding in the coop itself. The birds can kick and scratch through the weeds and trimmings to their satisfaction without old bits of plant matter getting mixed into whatever bedding might be in the coop itself. Meat scraps should probably be fed in trays which can be washed out after the meats are gone. That also ensures that the poultry owner can measure out a reasonable amount of scraps which the birds will quickly consume. Even if the owner wants to feed out a lot of scraps, they can be apportioned this way with almost zero waste or mess. Offal should also be fed wherever it's dry, so that any uneaten bits will dry out and break down before spoiling. The risk here is twofold. First, large piles of offal can draw all sorts of vermin from flies to rats and larger animals. Some of those larger animals wouldn't mind dining on the birds themselves, so we don't want to give them a reason to seek out our flock. Secondly, large piles of offal can create a stink problem. If that pile is out on the back 40 on some large farm, that's generally not a problem. If it's in the small backyard of a residential area, the neighbors are going to start to complain, or their dogs will dig under the fence and help themselves. Or both. While this scenario doesn't come up too frequently anymore, it has been enough of an issue in the past that it bears some mention. If a flock owner is dealing with large quantities of offal, a better solution may be to compost those remains. Or possibly, freeze them and portion them out in manageable amounts. Finally, the flock owner may want to check local ordinances regarding offal handling. Many areas now either recommend or require that offal be composted or carted off-site. It would be a shame to lose the ability to own birds because of mismanagement of this aspect of bird ownership.

Waterers are much easier to write about, because there's no concern about spoilage or smell. The only concerns with water are spillage, freezing and easy access. There are two basic types of waterers for birds – a reservoir type and a nipple type. The reservoir has a round cylinder body which holds the reservoir, and a tray at the bottom where the water pools in measured amounts. There are all sorts of sizes and various designs, but they all operate on the same principle. The reservoir can only be emptied through a small opening. If that opening is exposed to air, then bubbles of air can enter into the reservoir, and water can leave the reservoir. But if the opening is submerged underwater, the vacuum created within the otherwise closed reservoir prevents more water from escaping. That opening is in the tray itself. Thus the tray must be drawn down or empty before the opening is exposed to air; as soon as the tray has filled again with water, no more water comes out. It's really a very elegant design. Low tech and low maintenance for the owner, with near-perpetual fresh clean water for the birds. Two notes however. First, the tray base should be a few inches off the ground to keep the water clean, and the waterer should be checked every day to make sure it hasn't gone dry. Secondly, the waterer must be kept relatively level without major high or low points for the tray itself. If the tray is not level, such that the opening is continually exposed to air, the entire contents of the reservoir will empty into the tray and spill out on the other side. This can also happen if the waterer is knocked over. For this reason, many waterers are designed to be hung up from a chain or rope, such that the waterer's own weight will continually seek a level orientation.

The second type of waterer, namely the nipple waterer, is relatively new in the poultry world. But this style is gaining favor. Nipple waterers are a slender, hollow metal tube with a wide opening on one end, and a small ball or valve at the other end. The wide open end is inserted into a tank, bucket or other reservoir, far enough to get good access to the water within, while the rest of the tube sticks outside the reservoir. That far end drips just enough that poultry very quickly learn to peck at it, such that the valve or ball opens enough to let more water down. While there is a certain learning curve involved for the birds, most birds figure it out pretty quickly. This type of waterer works well when the flock owner has a lot of birds to water, because he/she can set up the nipples from any reservoir that serves the purpose – a five gallon bucket to a 55 gallon drum, or even larger if so desired. The flock owner can also install a float valve in that reservoir, since the valve doesn't depend on a vacuum as with the reservoir waterers described above.

Both types of waterers are prone to freezing closed, so folks in cold climates must have a procedure for watering their birds during below-freezing temperatures. The hanging waterers are generally small enough that they can be brought into the house and thawed out. In extremely cold weather, having two sets of waterers – one in service and one thawing out – is a nice way to guarantee good water access without a lot of chipping ice to provide it. For the nipple waterers, a tank heater can be inserted into the reservoir to keep that tank nice and warm. One caution with this method – if temperatures drop too far below freezing, the nipples themselves can freeze even when the tank is heated. So the flock owner will have to experiment with his/her own setup to see what works best.

## *Coop Furniture: Roosting Areas*

Chickens have some fairly particular preferences when it comes to their sleeping habits. First, they strongly prefer to be off the ground, as high as they can reasonably get. A sick or weak bird will occasionally roost on the ground, and broody hens will sleep right in the nest box if allowed to do so. But in general, poultry want to be up in the trees, whether natural or man-made.

This preference has traditionally been satisfied by providing roosts within the coop. Roosts can be made in a variety of ways and with different layouts, but they share some specific characteristics. First, the best roosting material is wood. Many folks have tried other materials, such as metal and PVC, for instance conduit. Unfortunately, the PVC is too slick for them to comfortably hang onto. Metal conduit has the same problem, with the additional problem of sometimes getting uncomfortably hot or cold in weather extremes. Wood branches or dowels are strongly preferred over both those materials. The wooden surface is rough enough for the birds to maintain a good grip, yet smooth enough that they don't cause sores on the birds' feet. Furthermore, wood doesn't conduct heat, so it stays comfortable regardless of the weather. Happily, wood also tends to be the cheapest material of the three. Branches can be used very nicely as roosts, as long as they're roughly 1.5" – 3" in diameter. Smaller and larger diameters can be used but smaller diameters won't reach as far without support. Diameters larger than 4" are unnecessarily heavy. That 2" to 3" diameter seems to work really nicely for all breeds of poultry, from bantams all the way up to turkeys.

Roost space is the length of roost which is needed for each bird. Many books will talk about minimum roost space allowances, so that each bird has sufficient space. Most common poultry breeds need 9" or so of roost space per bird. In other words, 6 birds of average size would need 6 x 9" = 54" or 4.5' of overall length. However, the birds also need room to get up to and then maneuver on that roost. For instance, if a coop needed to house six birds, a single roost of 4.5' high up against a wall might seem sufficient, but it wouldn't be. Most poultry can't vault straight up into the air and fly. That varies by the breed, with banties and small athletic birds being able to get up to relatively high roosts with ease. Larger birds such as the buff Orpingtons and Brahmas may only get a few feet in the air before sinking back down despite a feathery flurry of effort. That's why many roosts are a series of bars starting about 15" off the ground and stepping up two feet at a time. Birds can generally leap from one roost to the next, and thus work all their way to the top. If space is at a premium, a ladder of sorts can be created rather than taking up the entire width of the roost bar itself. Many different ladder configurations can and have been used, including a single beam with pegs or branches sticking out on each side, such that the bird can hop from branch to branch all the way up. In that instance, each branch needs to be substantial enough to withstand the weight of the bird when she lands and takes off.

Furthermore, the birds need to be able to move around a little from side to side, even after they've gotten up to that roost. Once everyone is settled in they'll pack into that minimum roost space. Yet they'll need "elbow room" on either side as birds are jumping up to the roost, turning around and otherwise getting settled.

Also on the topic of roost space, pecking order will definitely come into play with the strongest, boldest birds commanding the highest perch. They'll actually peck at other birds and drive them away. If the hen is particularly big or bold, she may create an empty space around her that could comfortably accommodate several more birds. For all these reasons, more roost space is preferable than the minimums listed. Allowing 12" to 15" of space per bird might seem luxurious, but that space will definitely get used, if not for roosting than for shuffling and fidgeting as the birds settle in each night.

Most coops have their roosts arranged in stair-step formation, with each roost about 2' above and 15" behind the one below it. That very nicely allows the top of the roost supports to be anchored to the wall or sill of the coop itself, while the bottom extends out away from the wall. The highest roost in that configuration will become prime real estate for the most dominant birds, with the lower roosts used to get up and down, and as roosts for the less dominant birds. Surprisingly, the birds may roost throughout the day, coming down to eat, to drink, or to lay eggs. They will snooze, preen and socialize up in the roosts if given the chance to do so. And in inclement weather, they'll seek out the roosts as if it was the end of the day. So when setting up roosts, be generous with roost spacing, and provide rugged materials. They'll get a lot of use.

## *Coop Furniture: Nest Boxes and Egg Laying*

If there's a single feature of chicken coops which allows for variation and creative thinking, it is nest boxes. Chickens can and will nest in all sorts of bizarre places – buckets laying on their side, planting pots, cardboard boxes, spaces between wall and piles of blankets, under the house, within clumps of tall grass, at the base of trees, and behind some old stacked lumber. Our egg laying hens have laid eggs in all those spaces, and a few others, over the years. They also lay eggs very cooperatively in more conventional nest boxes, conveniently right in the coop. Given the choice, they'll lay in all those locations and a few more I haven't thought of or found yet. So the conversation on nest boxes will be one part studious recitation of what works, and one part warning that they'll ultimately lay wherever they want.

Poultry are finding their way back to the yards, gardens and balconies of more and more homes these days, thanks to the eggs they produce. No more, and no less, than the good old chicken egg. If those eggs can't be easily laid, collected and used, chickens lose their value. As such, nest boxes are a crucial part of even the most basic chicken coop. Given the eggs' importance, their fragility and a hen's preferences for where to lay, nest boxes need some consideration.

First, let's talk about conventional nest box criteria. If someone were to design the ultimate nest box, it would have the following features:

- Approximately 18" to 24" long from front to back

- Approximately 12" to 15" wide from side to side

- Approximately 15" to 18" from top to bottom

- Have mostly solid sides, with a 12" by 12" front opening. Openings along the upper walls are a nice touch to provide for ventilation. A lip along the front edge is also nice to keep bedding in place.

- Bedding within the nest box of grass, leaves, shavings, hay, straw or similar.

- A certain amount of shading or semi-darkness.

- Moderate temperatures, between freezing and about 80F. Any lower and the eggs will freeze and crack. Any higher and fertile eggs will be able to start to incubate.

- Dry, dry dry. A nest box should provide shelter from rain and snow.

- Be kept clean. A nest box which has become fouled with mud, excrement or broken egg shards, will impart an off taste to any newly laid eggs.

- Above all else, be a quiet, relaxed, private place where a hen can retire and do her business without being interrupted.

Judging by the hundreds of hens we've had over the years, egg laying is a business they take very seriously. It's a regular biological process which they are compelled to complete, and they'll make do if they don't have access to something similar to the above. We have seen that a sudden loss in nesting areas will decrease, but definitely not eliminate, the urge to lay. We've also heard stories from close friends, about how hens in flooded out barns will safely roost up in the rafters, but then drown when they finally, irresistibly, feel compelled to come down to earth to lay. So it's an instinct that we have to honor, and work with. Happily, if we can provide nest boxes which meet most or all of the above criteria, they'll lay very nicely for us. And those eggs will be in good shape for us to collect and use.

Most hatchery catalogs and a large number of farm supply stores sell nest boxes of various types, and any of these will make perfectly suitable nest boxes, if the flock owner wants to pay the money. However, many folks probably already have materials at home which can be turned into perfectly suitable nest boxes. Here are items we've used in the past, and still use, as nest boxes:

- Old cat litter boxes with lids

- Cat carriers and dog carriers of all sizes

- Large plastic buckets with one flat side (for instance, those intended to rest flat against a stall wall), laying on that side.

- Planters and planting pots

- Cardboard boxes

- Wooden fruit crates

- Medium sized plastic tubs

While chickens can and will seek out such containers in which to lay, we can deliberately arrange our nest boxes in ways that are convenient for both our birds and ourselves. Within a coop, or immediately nearby, a row of nest boxes can be arranged on a stand or shelf anywhere from 2' to 3' off the ground. Being so arranged, the birds will check them out for several days, and then begin to lay. The main reason for having the nest boxes up on a stand is for our own convenience in collecting the eggs. Chickens will very happily nest in a ground-level nest box, but getting down on our knees to collect them every day gets really old really fast. That platform provides a much easier height for collection.

While any material can be used as a nest box for at least a short time, some materials have proven themselves to be durable and suitable over the passage of time. Of all the materials and items listed above, our favorites have become the dog and cat carriers. Being plastic, they are lightweight yet rugged, easy to wash out, hygienic, and most carriers have cutouts or grates high along the sides and/or back which permit light and air while still protecting the contents. Old litter boxes with a fitted cover are our second choice, for most of the same reasons.

One well established and helpful rule of thumb for nest boxes is to allow one box for every four laying hens in the flock. If the flock has only four hens, a second nest box would still be a good idea simply to give the hens an alternative place to lay, if the first nest box is occupied.

Another rule of thumb is that birds generally require 14 hours or more of daylight to lay. Their laying hormones are governed by the number of hours of light and dark during any given 24 hour period. During winter, or if the coop is in deep shade, the flock may not get enough daylight and their egg production will drop dramatically. Many birds will also start to molt during short-day periods of time such as winter. For these two reasons, many (if not most) commercial egg farms have their flocks in lighted barns during the winter, such that they always get at least 14 hours of light. This keeps them laying right through the winter, and it delays their molting process. For smaller flock owners, a few simple lights in the coop, set up on timers, can help ensure that the flocks continue to lay through the winter. The further away from the equator folks live, the more important this becomes. We have southern friends who don't bother to supplement their birds and young birds will just keep right on laying through the winter. But up here, a mere hundred miles south of the Canadian border, we have to start supplementing our birds before the first day of autumn if we want them to lay through the winter. Alternately, some folks deliberately let the light levels fall, or they keep on the supplemental light until a fixed date, at which time they turn off the supplemental lights for a period of about two months (often, the December/January timeframe when days are already shortest). This management strategy does two things – it ensures solid egg production right up until that date, and it synchronizes the entire flock such that the flock stops laying, and starts molting, all at the same time. This offers egg gatherers a nice break from the daily task of gathering and washing eggs, and it gives the birds a chance to molt their old feathers and grow in new feathers.

When the birds start laying again after their molt, they usually won't lay at their previous production levels. For commercial egg laying farms, they retire birds during their first molt because they never come back to their previous productivity. For hobbyists and small scale flock owners, those birds can continue to lay for some time before they stop laying altogether. Even five or six year old birds can sometimes still produce an egg or two a week; if they are a favorite family pet, it may not matter that they're laying at all. It is ultimately up to the flock owner to decide when a bird's egg laying career is truly and finally over.

A note on selling eggs. Many folks want to acquire a few hens, such that they can gather eggs for their own use. As of this writing, if the hens are permitted by municipal law, then the owner's use of their eggs for personal consumption is permitted as well. However, the moment folks want to start selling those eggs, even to friends and neighbors, a number of additional rules and regulations kick in. Some locations have fairly lenient rules in this regard, requiring only that the eggs be washed of obvious dirt and contamination, and be refrigerated prior to sale. Other areas have extensive rules about how the eggs are to be washed, what temps they're kept in, and how long they can be kept prior to sale. Even details such as where the wash station is located, is a matter of regulation. Some cities, counties and states specifically forbid kitchen prep because that location is too open to other normal family activities. Flock owners in those locations would need a separate wash station in a separate building. That topic goes well beyond this guide, but flock owners who wish to sell eggs should look into their local requirements. All the comments I made earlier about renegade chicken flocks apply here too.

## *Coop Furniture: Brooders and brooding*

Many folks won't want to tackle this aspect of poultry raising, particularly in areas where roosters are prohibited. Yet for those households and small farms which want to raise their own birds over time, brooding becomes the next step in poultry raising. This is another topic that goes well beyond the scope of this guide, and a number of publications and books have already been written on the topic. Here I'll only mention some basics to keep in mind while designing the coop.

Chicks are generally raised in one of two ways – naturally with a hen, or artificially with an incubator and brooder. If the former, the hen will build a nest of eggs, usually in some hidden place. If a flock owner gathers eggs every day and finds the same hen still in the same nest, and she apparently spent the night in that nest, she's gone broody. If the flock has no rooster, those eggs won't be fertile and she won't be able to raise any chicks from them, despite her instincts to incubate. In that instance, the flock owner's best option is to remove her from the nest (she'll protest loudly) and try to keep her from using it for the next few days. If the flock owner can get her to roost normally overnight for a few nights, that will usually trigger hormonal changes within her body such that she doesn't feel the drive to brood anymore. She'll start laying eggs normally again and return to roosting at night. If, however, the flock does have a rooster, some of those eggs may be fertile, and the flock owner will have a decision to make. The owner can leave her in the nest and let her try to raise whatever eggs hatch out. Or remove her, remove the eggs, test to see if any of the eggs are fertile, and incubate any that seem clean and whole and fertile. Be advised that some breeds of chicken go broody very easily; Buff Orpingtons are renowned for their broody behaviors. Other breeds have almost no broody tendencies. If a flock owner definitely wants, or doesn't want that behavior, selecting breeds which do or do not go broody. That choice will save the owner some grief.

If a flock owner does want to raise homegrown chicks, allowing for space within the chicken coop is an excellent way to do so. The owner won't want the young chicks mixed in with the adult birds right away. Size matters in a chicken flock and young, small birds get picked on, and pecked on, sometimes to the point of serious injury or death. So a flock owner will have to have a small pen, cage, tank or other sequestered area away from the main flock. He or she will also need a source of supplemental heat for the chicks. They need temps of nearly 100F their first week of life, decreasing by about 2F with each passing week. They'll need their own feeder and waterer so that they're not competing with the older birds for either. And they'll need protection from drafts. A 3' diameter or larger metal livestock tank is a great way to brood a batch of chicks, up to 25 at a time, within a larger coop. Most of what they need can be contained within a tank of that size, or can be hung above. So if that is of interest, leave room for that in the coop designs.

## *The Human Factor* –

## *Working Conditions, Efficiency, Sanitation*

Regardless of the number of birds, size of coop or other design features, most flock owners will need to go check on their flocks at least once a day, if for no other reason than to ensure everything is as it should be. That means the flock owner will want that coop to be as decent a working area as possible. That might sound like a luxury, but it's actually very practical. The more efficiently the coop is laid out and the more efficiently a flock owner can work in it, the faster those chores will be and the easier it will be to keep the coop in good working condition. Some careful considerations in advance, will save a world of effort later.

First, consider the working conditions. If the flock is small (say, less than a dozen birds), the coop may be sized such that only the birds will be walking around inside. Many small flock owners have coops which would fit on the kitchen table. That is a very efficient use of building materials. However, the design for that ultra-small coop needs a lot of consideration. First, how will the eggs be gathered? The prospective flock owner doesn't want to have to crawl into the thing just to get the eggs. An outside hatch door, with a simple latch, can provide very easy access without even bending over. Second, how will the coop be cleaned out? This will be determined in part by the size, and in part by whether the coop is stationary or mobile. If the coop is a hutch-type, i.e. too small to walk into (and usually up on a stand, or on wheels), the bottom of the coop should have a mesh floor so that manure can pass right through it to the ground beneath. That dramatically reduces the need to clean the coop. With this setup, some owners would merely clean the coop in between batches of birds, on a roughly every-other-year basis. On a related note, how will the nest boxes be cleaned out? While a simple hatch at the back can be used to gather eggs, it can also be used for cleaning purposes if it's large enough.

If the coop is large enough to walk into, the flock owner has a few more things to consider. Human doorways should be at least 2' across, but 30" across is better. While our bodies don't need that much space per se, we'll need it if we're carrying something (like a feeder or waterer). In addition to a nice big door, providing for cross-ventilation is good not only for the birds but also for the person working in the coop. Even a clean coop will have a certain smell to it, and being able to open up the coop on nice days to air it out is a distinct advantage. During hot, calm weather, having that cross-breeze can be a life safer for the birds. Headspace becomes important too. Yes, we can duck under a low ceiling if/when needed. But I can testify from lots of personal experience, that bonking my head on some low ceiling rafter gets really old, really fast. Third, there's maneuvering room. Leave at least a 2' walkway (30" is better here too) between things like the nest boxes and the roosts, the roosts and the feeder/waterer, etc. It's perfectly acceptable to have a single walkway down the middle of the coop, with nest boxes and feeders/waterers on one side, and roosts on the other. Or some arrangement where all the furniture is against the various walls, or the walkway is against a wall. Just make sure to leave enough space to get in there and comfortably move around, particularly while carrying something. Spilling a full waterer or a full feeder, because of running into the roost again, is a very good way to start the morning off on the wrong foot. And make sure the walkway doesn't pass directly underneath the roosts. If it does, the birds can and will eventually poop on unfortunate owner as he or she is walking along doing something after hours in the coop. Sufficient light, via either a ceiling fixture and/or windows, helps a lot as well. We've already touched on the importance of light for egg laying and molting but we'll get into it in more detail a little later. For now, just know that it's important enough to design in advance. All of these items can be provided easily enough if designed right into the coop from the beginning. They are more expensive to add after the coop is already built.

Let's take a moment to talk about efficiency. This may not seem important, but I've yet to meet the person who has so much time and money that they don't mind wasting both. And a poorly laid out coop will waste both. Savings for a small flock may not add up to much, but a penny saved is still better than a penny wasted. Some simple ideas: store the feed nearby, in metal cans or bins, so that they are easily accessible while still being protected from weather, spoilage and rodent damage. Have the feeder set up so that you can either easily carry the feeder to the feed supply, or vice versa. Some feeders are designed to be filled each day with a measured amount of feed, to be eaten quickly. These feeders usually minimize spillage and thus minimize waste. Conveniently they also minimize rodent access to feed. Other feeders are designed to be filled every few days, and they ensure that feed is available all day long to the birds. These feeders can be a real benefit if rodents aren't a problem. However, having feed available all day like this will draw rodents like a magnet, so the flock owner will either need other rodent prevention/elimination features in place, or that feed will end up feeding a very fat population of rats and mice. We'll talk more about this aspect of things under Vermin Control.

When feeding, particularly if a flock owner is using the fill-each-day type feeders, it helps to measure out the amounts being fed so that the flock is consuming the same amount each day. This gives owners a very good indicator of how much feed the flock is using. Furthermore, the owner will be able to dial in the amount to feed per day so that all the birds get enough, without a lot of waste or spillage. Feed can be measured out in something as simple as a used tin can, plastic or metal coffee can, small feed supplement tub, or a dedicated feed scoop with volume measurements etched into the side. I use a heavy duty plastic, quart-sized measuring scoop intended for restaurant use.

Waterers aren't typically as big a deal as feeders, except in wintertime when frozen water is an issue. For those folks who will have to deal with occasionally or regularly frozen waterers, there are some options. First, the metal waterers can have a heater placed underneath them to keep them warm. But that requires an electrical outlet somewhere in the coop, which then must be shielded against being pecked at or otherwise messed with by the birds. If the outlet projects even a small distance away from the wall, the birds will try to use it as a landing pad or perch. So it'll need to be flush with the wall, or otherwise mounted in such a way that they can't land on it. The same goes with any wiring from the fuse box to the outlet, and from the outlet to the waterer heater. Another option for frozen waterers is to have two waterers at any given time – one thawing in the house, and one being used in the coop. That's what we do, and it works OK. I'd rather have the heater, but I don't currently have a safe way to set that up. Folks can also use the relatively new watering nipples with a big tank, and then put a tank heater right into the tank. The heater per se is thus protected from the birds by being submerged, and the sheer volume of the tank will often keep the nipples from freezing unless the temps really go down. For those folks who have temperatures which go well below freezing and stay there (say, 20F or less), the tank heater alone may not be sufficient and the flock will still need to have some alternatives ready to go. Some combination of large tank and nipples for three-season use, and portable waterers which can be thawed out inside, is a workable compromise.

Now let's talk about sanitation. No one likes to work in filth, and chicken coops can get filthy pretty quickly if allowed to do so. If a coop has a solid floor, the birds' manure is going to pile up very quickly, and become very smelly. A small coop up on stilts or wheels can have a floor made of hardware cloth or other metal mesh, to allow the manure to pass through. This is a fantastic design if the owner wants, for instance, to park the birds over a certain spot in the garden, so that they can fertilize it. For larger coops, the owner will need to consider both the ground that the coop is on, the bedding they have, the flooring (if any), and how the owner wants to use or dispose of all that manure. We'll cover all this under Bedding and Waste Management.

## *Protection From Weather*

Chickens are quite hardy folk, and they do quite well in most weather conditions. Situations where they do need protection are:
- High temperature days (90F and above)
- Rain/Snow
- Cold and Wind

Poultry generally enjoy basking in the sun, but when temps go much above 85F, they start seeking shade. If the temps go above 90F, the birds will stand heads down, with their wings partially spread, and they may begin panting. As we noted above with coop designs, having a well ventilated coop can be important during high temperature days so that the birds have a place to cool down. If they can get up on the roosts, into the moving air, open those wings and let some of that body heat disperse, they'll be much happier than if they are in still air.

Conversely, when the weather turns cold and nasty, it's nice to be able to give them not only a roof to protect against rain and snow, but also walls or windbreaks to take shelter against wind-driven moisture, and cold drafts that would pull body heat out of them. A bird's main insulation is their feathering, and wind will ruffle through those feathers and leave them almost unprotected during times when they need that protection the most.

For super-efficient housing, the single best place then to put roof and walls is over the roost area, where the birds retire for the night and where they'd instinctively take shelter during bad weather. They can handle extreme cold, even below zero temps, if they can stay dry and out of the wind. Their feed and water supplies should be similarly protected, not only to keep rain and snow from getting into the feed, but also to give them a protected area while they eat.

One commonly overlooked issue in this topic, is mud. When birds are living either on dirt floors, or they spend time in a yard exposed to the elements, they can and do track mud all over the place. They'll track mud into the nest boxes, such that the eggs and bedding turn muddy. They'll track mud up onto the roosts and then roost on their muddy feet, such that their abdominal feathers get muddy and the roosts get caked with mud. For this reason, it's generally a good idea to try to keep them off muddy ground as much as possible. That may or may not be feasible depending on the location, the yard and the weather. But a small dry exercise area, full of sand or wood shavings, is better for them than a larger yard that has turned into a mudpit. In some climates this won't be an issue, but for us and for folks in cold rainy climates, or where there is a monsoon season, it can become a real issue. Consider having a "wet weather" exercise yard that is kept covered and dry, for these types of days when their normal yard is just too wet and sloppy. The flock owner will save him/herself a lot of egg cleaning and coop cleaning time as a result. And those birds will be healthier too.

## *Vermin Control*

One issue of utmost concern to many residential flock owners, the public health agencies that sometimes regulate urban flocks, and farm producers in general, is vermin control. That will usually consist of rats, but it can range from tiny little field mice nibbling on bits of spilled feed, all the way up to bears knocking over and raiding the feed bins. Some flock owners would include predators in this list of vermin. We'll consider predators next, in the context of those animals which seek to eat the birds themselves. But for now let's focus on those animals which don't eat the birds per se, but are attracted to the bird feed. For the sake of discussion, I'll focus on rats and mice but include other suggestions which pertain more to larger animals, as needed.

Rats and mice in particular, and other animals in lesser numbers, are attracted to bird feed like magnets. Who wouldn't be? Bird feed is high protein, high vitamins/minerals, easy to find (right there in the convenient feeder), doesn't run away, is fresh (at least it should be!), and it's replenished every day. For rats and mice, it's basically like a self-serve cafeteria just opened for business. While a stray field mouse or two may seem perfectly innocent, their numbers can increase with astonishing speed and become a real problem. They foul the food they eat by peeing into it, they can track in disease which the birds can then get, and they will eat the flock owner out of house and home if given half a chance. So controlling vermin populations will, sooner or later, become an issue for any flock owner.

Happily, there are a variety of ways to control, or at least minimize, rodent access to the coop. As we've discussed before, including these details right up front will make construction cheaper than adding them after the fact. Nevertheless, many folks are faced with an existing and/or worsening rodent problem, so they'll need to consider how to retrofit existing coops, feed areas and yards. Let's take a look at some of the options.

First, rodents are there for the free meals. They are not initially attracted by the birds themselves, and rarely ever attack the birds. They will attack chicks if given the chance, and they will consume a bird that is either incapacitated, or which has already died. In short, they'll go for whatever is the easiest meal. So a lot of our vermin control is to make that meal as hard to get as possible.

Let's start with the feed storage. For small flock owners, it's tempting to bring home feed in the 25# or 50# bags, and just open a corner of the bag to feed from. This presents a few problems. First, rats, mice and other uninvited guests can get into the feed not only through that opening, but they can chew right through the bag itself. Since most pest species are fairly shy by nature, they won't start where the flock owner can see the gnaw hole, but rather they'll start at the bottom, typically along the side of the bag that is conveniently leaning against a wall. And they won't be tidy in their invasion. They'll nibble a little here, then spill some more and then nibble a little there, then spill even more and nibble again. If the problem gets bad enough, and/or their populations get high enough, they can turn a single bag of feed into a scattered expensive mess, in a single night. Raccoons would leave a bag of unprotected feed scattered all over the yard. Bears would simply eat the bag along with the feed. For this reason alone, feed should never be stored in the original bags. There are a few other reasons feed shouldn't be stored in the original bag: water damage, loss of nutrient value, spoilage, etc. But vermin damage is usually at the top of the reasons list.

So, if folks shouldn't store feed in the bag, what storage options are there? The simplest, and by far the most common, is to use a galvanized metal trash can or similar metal container. A typical 30 gallon metal trash can, with closely fitted lid, will hold 150# of feed. Such cans are nearly rodent-proof because the rodents can't chew through the metal (they can chew through heavy plastic trash cans and wooden bins, although it would take awhile). For those in raccoon country, that tightly fitted lid needs to be backed up with a solid latch that can't be fiddled with or undone by dexterous animal paws. For those folks who are in either bear country or feral pig country (or who have other animals on the property that might like to snarf down a bunch of grain), putting the feed can in a locked room, with a sturdy door, is the only sure way to keep all possible animal raiders out of the feed. For folks who own goats, sheep, cows and horses, that strategy can actually be a life saver. While those animals can safely eat non-medicated chicken feed without harm, they will gleefully gorge on any unattended bag. That gorging can and has caused death due to impaction, bloat, colic, and other types of digestive upset. In general, it's always safest to lock up the feed in a metal bin, and lock up the bin in a secure room. Everyone is safer that way, including the flock owner's wallet.

Now that we've talked about how to protect the feed supply in general, let's talk about how to protect the feed supply right in the chicken coop. Flock owners have a few different ways to do this, and the best option(s) will depend on the size of the flock, the feeding schedule, and the design of the coop and yard.

One strategy for minimizing rodent issues, is to minimize the amount of feed that's available at any given time. As we saw in the feeders discussion, flock owners can choose two different feeder designs – a bulk feeder which is refilled when empty, and always has feed out for the birds, or a tray feeder which is filled with a measured amount, intended to be eaten in one feeding session, then refilled the next day. For busy folks who rely on bulk feeders so that they don't have to feed every day, rodent control can be a real challenge because that feed is just sitting there, waiting for some clever rodent to figure out a way in. So the first suggestion is to use tray feeders and measured amounts of feed. By limiting the feed to what the birds will actually eat each day, the flock owner can minimize the amount of feed lost to the rodents. The birds will normally compete pretty well against the rodents, who will wait until the birds are done or nearly done before approaching the feeder. If there's nothing left for them to eat, that's what we're aiming for.

A second strategy which works with varying degrees of success, is to use barn cats, dogs, or even very possessive roosters and large birds to drive the rodents away. This approach is very common, but it's not particularly reliable. We've had barn cats now for a very long time, and they are rather fickle in their hunting. Some cats have no hunting skills or ambition whatsoever. They'll watch rodents scamper around without ever lifting a paw to stop them. Other cats are veritable hunting machines, and will strike with stunning speed and dexterity. But be aware that even a good feline hunter, will occasionally be tired and/or full of rodent from the last hunt, leaving the next rodent free and clear to go claim a meal. We have been fortunate here to have a line of cats that started with a mom cat who was one of those amazing rat killing machines. One of her daughters, at least two of her sons, and two grandsons inherited that skill. Another daughter did not. What has worked decently well for us is to have several cats of at least average hunting ability/ambition, circulating through the chicken barn at all times. If rat populations are low, or the cat population is high AND are proven hunters, that combination is often enough to keep rodent populations minimized. In our case, we unfortunately let the rodent populations build up so high that the cats can't keep up. So while they can be fed very nicely from the population, they aren't controlling the population. But my goal this winter is to upgrade our facilities to what I'll describe next. That way, the cats can patrol the chicken coop and merely guard against the occasional scouting party, rather than dealing with a well entrenched rodent population.

A third strategy takes a lot more effort, but may be very much worthwhile in the long run. Namely, build a rodent-proof coop and/or yard area. It sounds like Mission Impossible, but it's not. One of the very best chicken facilities I've ever seen, had a very simple but amazingly successful design in this regard. First, both the coop and the exercise yard were on a bed of sand. No particular kind of sand, but leveled and flat and contained by railroad ties. I believe the exercise yard measured about 50' long by maybe 15' wide. Second, the yard itself was enclosed in 1" chicken wire, but the bottom 24" was also enclosed in ½" hardware cloth, with a hotwire running just along the top of the hardware cloth on the outside of the pen. That combination means that rodents couldn't get through the lowest sections of fencing because of the hardware cloth, and they couldn't climb the chicken wire because of the hotwire deterrent. Third, the exercise run was also closed at the top with bird netting, so the birds couldn't fly out (and avian predators couldn't get in). Fourth, the chicken coop was small enough to be located right within the yard. Since the yard was already rodent proof, the coop itself was fairly simple – a floor of plywood on joists, with stud construction walls, a shed roof, and relatively conventional roosts, human door, chicken door, nest boxes, etc. Since the whole thing was on sand, rodents couldn't dig their way underneath (a rodent can't dig tunnels in sand). Their chicken yard was the only chicken yard I've ever seen, which was completely and totally rodent-free. And just as extra insurance, their farm cats circulated freely around the exercise yard. That is the design we're moving towards here.

With that design, the coop itself can be almost any design that works for the number of birds, the size of the yard, etc. In their case, they had several yards of this size, each holding approximately 25 birds. The bulk of the cost would have come from the sand pad – that ground would need to be leveled, then the railroad ties brought in as retaining walls, then sufficient sand brought in to fill the yard. The yard fencing would have been another slightly higher cost but most poultry yards must be fenced that way anyway, to provide some measure of predator control (which we'll talk about next). By the time the sand and the exercise yard fencing had been provided, the coop itself could have been a shack made from free pallets and still provided sufficient protection. It happened to be a handsome 10x10 stud construction coop, but that's because the owner wanted it to look good. Talking to the owner of that fine setup, the only complaint he had about the design was actually the occasional snow load on the bird netting cover for the exercise yard. He wanted to rig up a way to draw that netting aside, such that the snow would simple fall into the yard and melt away into the sand, rather than sticking to and piling up on the bird netting. For folks in high snow areas, that would be a serious concern. For folks in no-snow areas, that wouldn't be any concern at all. From a rodent's point of view, that overall design turned that whole area into Fort Knox – the treasures within were simply too well protected.

One final category of vermin that we don't have here, but I know a lot of other folks struggle with, is snakes. Snakes won't generally eat chickens, but they can and do eat eggs. There are a variety of ways to deal with snakes, and a lot of folks are quite happy to simply kill them and be done with it. However, I would caution folks that many snakes are quite beneficial, eating animals such as the very rodents that we've already talked about trying to control. So killing that big snake that just started eating those precious chicken eggs, might be solving one problem while creating another. The better solution would be to ensure the snake can't get into the coop to eat the eggs. The above-described rodent-proof yard would be a good defense against most types of snakes. Snakes will try to squeeze through very small openings or will use existing rodent tunnels to gain entrance. If the coop is on sand, that means no tunnels. If the fence has the ½" hardware cloth at the bottom, that means not entry through the base of the fence. Now some snakes can climb trees and thus may be able to either drop in over the top of the fence, or could potentially climb the fence itself. Here's where the specifics of the location will come into play. For known tree-climbing snakes, simply eliminating branches and brush from near the coop, may very nicely eliminate that point of entry. For those who can climb fencing, more rigorous measures may be needed. Either a more substantial roof all the way around/over the exercise area, or some barrier midway up the fence to prevent the snakes from reaching the top. I don't happen to know if hotwire works on snakes; I suspect it wouldn't work very well. It won't work at all if they're not in contact with the ground. For those folks in parts of the country which have climbing snakes, the best solution might be to check with a variety of nearby flock owners and see what works for them.

## *Predator Control*

Now that we've seen how to protect birds against those who would eat their feed, let's talk about how to protect birds from those who would eat the birds themselves. There's a saying amongst chicken owners, a somewhat discouraging saying, that Chicken is on everyone's menu. That means that there's probably not a predator alive that would turn down a free chicken dinner. And by free, I mean "easily obtained." The flock owner's job is to make that meal as expensive as possible. Happily, there are a lot of ways to do that.

Let's talk for a moment about all the creatures that might eat chickens, because they all have slightly different ways of sneaking in and claiming a bird. First, there are daytime avian predators – hawks and eagles. These predators will sometimes dine on full grown birds, but they prefer chicks and younger pullets. They will often perch in a nearby tree and watch the coop for awhile, in effect "casing the joint", to learn the typical habits and schedule for the flock within. Once they've learned that flock's routine, they'll wait for their opportunity and make their move. They can strike so fast that if the flock owner wasn't there to see it and hear the commotion, he/she might never know they were there. They generally take the entire bird such that there isn't any evidence of predation unless the owner starts counting birds and come up a few birds short.

Nighttime avian predators, namely owls, are not as common but they do occur. If chickens are allowed to roost out in the open, for instance on exposed branches, larger owl species may prey on them. If the birds roost under a roof, this risk is essentially eliminated.

Daytime terrestrial predators are many and varied. Dogs and wild members of the dog family are notorious chicken killers. If accurate predation numbers could be added up of all chickens killed nationwide by predation, I would wager that loose neighborhood dogs, coyotes, foxes and even livestock guardian dogs would be the single biggest group of killers. They have the size, they have the speed, they have the dexterity to jump into the air when the birds take to panicked fluttering flight, and they certainly have the appetite. Even in town, a combination of loose neighborhood dogs, and/or urban coyotes (which are growing in numbers every year), can claim a tremendous number of chickens. They will dig through and under fencing, gnaw or scratch their way through wooden siding, and jump into the air to snatch birds off of roosts. Even the beloved family Golden Retriever, who never harmed a single hair on a kitten's head, may treat chickens completely differently. There's something about either the feathers, and/or the birds' wobbly gait, which brings a hunting gleam to many a dog's eye. Never ever trust a dog around birds until and unless that dog has been proven to be trustworthy with birds – being trustworthy with other livestock doesn't count. We have four livestock guardian dogs that are all trustworthy with our four-legged livestock, but only one of them is trustworthy with the birds and that's because we trained him to be that way. Better yet, simply don't allow the dogs access. And then everyone will sleep better.

Domestic housecats and neighborhood feral cats actually do pretty well with birds – the birds are too big for most cats to want to mess with, at least in our experience, and cat patrols will help limit the rodent population. We've never had Banties, and a big kitty living near a yard full of Banties might be a poor combination. But most other chicken varieties do OK with cats. Also, cats will gleefully eat chicks if given the chance to do so. Anyone raising chicks will need to provide special protection if the chicks are to live in a coop guarded/patrolled by cats. Even if the flock owner don't have cats of his/her own, never underestimate the presence of, and interest of, neighborhood feral cats. They'll come out of the woodwork if there are chicks out and about in the yard. Larger wild cats, such as bobcats and cougars, will take the occasional bird ranging out on pasture, but these big cats are usually fairly shy and won't come into well lit areas. However, if the property has a lot of tree cover or brush cover, they can and will climb trees, and/or hide under/behind brush, very close to human habitation. They'll drop or leap into the yard, claim their prize, then leap out again and be gone. I don't hear about big cat strikes against flocks as often as I do about dogs getting into flocks, but it can and does happen. If a flock owner is in an area with known big cat populations, keep brush away from the poultry yards and trim back any branches near or hanging over the yard. Having a roughly 10' to 15' clear perimeter around the chicken pen will prevent most big cat predation.

Aside from dogs and dog-family predators, raccoons, skunks and possums are the last major trio of terrestrial predators that a flock owner needs to worry about. While possums are uncommon or unheard of in some parts of the country, raccoons and skunks occur just about everywhere, even in cities. All three operate in much the same way. They have dexterous enough front paws that some folks simply call them hands. They don't have opposable thumbs like we do, but they may as well for the types of latches they can manipulate. They can pull apart many types of seams and joints, which is one of their main methods of entry. What I mean by that is they can pry apart the spot where to fence panels come together. They can pry apart the corner where two plywood panels or boards come together. They can pry open doors and windows and hatches. They can grasp and wiggle metal sliding doors and metal trash cans until they work a moveable piece enough out of the way to gain entry. And once inside, they can devastate a chicken coop. They will kill chickens for the fun of it. They'll tear them apart and never eat them. Or they'll eat some parts and leave others. Additionally, they'll go after the eggs. They'll eat them or smash them or both. In the case of these invaders, the flock owner will often see the evidence of their visit, but that owner might be stumped on how they got in. Dogs don't close the holes they made on the way through, but possum, skunk and raccoon are notorious for leaving only dead bodies or portions thereof, without a trace of how they got in.

The final category of chicken predators includes animals which aren't that common, and some folks will never see them. Yet they should be considered and a coop designed to keep them out. Namely, weasels, mink, ermine and wolverines. These predators are much like raccoons, skunk and possums. They can get in through seemingly impossible entrance points – tiny little gaps, electrical penetrations, knot holes in boards, etc. Wolverines are by far the largest of the group, and should be considered small bears in terms of their strength and potential for sheer damage. Thankfully, wolverines are also limited to quite rural northern or mountainous locations, so most folks won't have to deal with them. But weasels, mink and ermine are much more common. These wild relatives of ferrets are voracious carnivores and will gleefully eat anything they can catch. They'll also take on and win against animals much larger than they are.

So now that we've scared ourselves into a tither, thinking that the whole world is out to get our chickens, let's talk about how to protect them. Thankfully, most flocks, most of the time, won't have to deal with all of these predators. They'll only have to deal with one or two of them, periodically. Conveniently, many predator control methods which work for one species, will work for many others.

The first and by far most common protection against predators is a combination of good fencing and secure coops. Let's talk about fencing first. As we saw with the vermin control, a cleverly designed fence can go a very long way towards protecting the flock. For most predators, the only fencing variable that changes is the size of the opening. For instance, dogs and raccoons can't get through 2x4 openings in a woven field fence. But possum, small raccoons/skunks and any member of the weasel family could very easily get through that same space. Also, the type of fencing is important. Chicken wire is very familiar to most folks, and its 1" opening size will keep a lot of predators out, for a short time. Unfortunately, the wire is so thin that it either rusts through, or can simply be pulled apart, quite easily. Additionally, the way that chicken wire is manufactured, is actually fairly complex. The wire is spun into the familiar hexagon pattern, while the longitudinal wires are woven into it t the same time. I have seen enough instances where the assembly line must have been out of adjustment just enough such that the wires didn't snug against each other. It looked like a solid fence but a lazy overweight pug could have simply leaned on it and fallen through it to the other side. That is not good predator-proof fencing.

A nearly bomb-proof type of fencing for chickens is called hardware cloth. It comes in either ¼" or ½" openings, which even a mouse can't get through. Unfortunately, it only comes in rolls that are at most 30" wide, so a flock owner would need a lot of it to surround even a small exercise yard. It's also relatively expensive, when compared to other types of fencing. As a result, hardware cloth is best used for those places such as flooring, or at the base of other types of fencing, where access must absolutely positively be most restricted. But the diligent flock owner doesn't need to cover a tremendous amount of area – just those areas most prone to unauthorized entry.

A third option for fencing is less expensive and quite a bit easier to work with – a woven wire fence with 2" openings. I should stop here and explain two fundamental types of fencing: welded field fence, and woven field fence. Welded, as the name implies, has a multitude of small welds where the horizontal and vertical strands or wires overlap. Woven wire means that the horizontal and vertical wires are physically joined by either being woven around each other, or by having another small section of wire woven around the joint. Welded wire is dramatically cheaper than woven wire of the same dimensions and size openings, because it's quite a bit easier to make that spot weld at each joint, than to wind the strands around each other, and/or have another piece of wire wrapped around each joint. However, many types of welded wire have such a small weld at each joint, that any pressure on the fence breaks the weld. At that point, the flock owner has a fence of wire spaghetti that isn't connected to anything, and anyone could go right through it. Other types of welded wire are extremely thick and durable, such that it would take a lot of force to break those welds. The rule of thumb is that the easier it is to flex the fence panel, the easier those welds will pop free. If the fence is fairly rigid, and the gauge of the wire is 16g or larger, the fence panel will be heavy enough to withstand a certain amount of abuse. Lightweight welded wire would thus be a good choice for a fence that doesn't expect to get a lot of pressures on it – i.e., animals won't be leaning against it or rubbing against it or trying to claw their way through it. Heavy welded wire, such as cattle panels, hog panels and the rigid panels used in dog kennel runs, is very durable. Woven wire is also considered very durable.

Back to our conversations about fencing for birds. A woven wire fencing material with 2" openings is a good middle-of-the-road option for poultry in terms of expense versus durability. I have used a woven fencing product which is 48" tall, 50' long and with 2" openings, throughout our operation. Even after years of use (and abuse) it's still standing. By comparison, a slightly less expensive welded wire fence fabric, 60" tall and 100' long, with 2" x 4" openings, worked well for the birds along most of its distance but didn't stand up well to heavy use such as at gate locations, where I was coming and going multiple times per day. Had I reinforced that gate a little better with rigid framing, it would have held up longer. However, another disadvantage is the 2" by 4" opening size. Many kinds of small predators, such as small possum and vermin such as rats, can walk through that spacing with ease. When I redo my poultry yards, I'll be using the much smaller 2" woven fence panel for the whole thing.

One of the issues which some, but not all, chicken owners will face, is that of birds flying over relatively low fencing. We've had a lot of different breeds, ranging from small, near-banty sizes, all the way up to some of the giant breeds. In general, the larger they are, the less inclined they are to fly. The smaller they are, the more easily they can (and will) take to wing and fly up or over whatever buildings and fencing is installed. Meat-type birds as a rule won't fly. But many egg-layers will. We currently have a bloodline of Buff Orpington crosses, which are normally fairly large birds, who can and do enjoy flying enough that we find them on top of our chicken coop. They'll easily top a 6' fence. A smaller breed we have called the Hamburg, is equally agile and even more prone to exploring. They are so inquisitive, we've never found a way to keep them inside the fences, other than to put bird netting over the top. Yet other birds we have, both smaller and larger, never test the fence at all. So any poultry fencing design will need to at least consider the possibility that birds will fly over it unless contained with a roof of some kind. Fencing can't do much to protect the birds from predators, if they won't stay inside it.

Another thing that folks can do to protect birds against predators, is to keep dogs. We have livestock guardian dogs here that live outside 24/7 either with or near our livestock pens, and that proximity has virtually eliminated our terrestrial predation issue. Not all dogs will have that effect; we had a house Labrador for 15 years who lived inside but spent a fair amount of time outside. Predators ranging from possum to raccoon to coyote would come and go on the property with ease, merely waiting for her to go inside and then moving through. And yes, they did claim some chicken victims along the way. However, as soon as we got our first livestock guardian dog AND stationed that dog near our poultry pens, predation stopped and has never returned. In our case the livestock guardian dogs don't live with the birds themselves; ironically most of our dogs would actively try to catch the birds themselves if they could. However, their mere presence is enough to drive off any visiting terrestrial predator.

A third thing that can help protect against predators, is to keep clear buffer zones between the bird pens and nearby trees, brush or heavy grass. Most predators won't merely stroll up to a pen in the open, and try to get in. At least not at first. They'll look for the best concealment point, so that they are exposed as little as possible. And that's where they'll try to penetrate the fence or building. If a poultry pen has a clear buffer around it, AND that buffer is frequently patrolled by either dogs or people, that can discourage a predator from trying to gain access.

A fourth protection against predators, is electrified fencing (more commonly known as hotwire). One or two well-positioned strands can do a lot to discourage predators from penetrating a fence. That hotwire is particularly effective if it keeps them from exploiting a weakness such as where two fence panels or walls come together. While hotwire might not be particularly cost effective for a small flock, it can make a very nice difference for protecting yards from a variety of predators. I should mention that this type of electric fencing features one or more hotwires located above-ground, with the intention that predators will hit the hotwire while checking out the fence. A hotwire fence is not to be confused with so-called invisible fencing, where the wire is buried and the dog must wear a special collar before getting the shock. Invisible fencing has zero effectiveness against anything not wearing that collar. Sadly, it often has little effectiveness for dogs even when they are wearing the collar. Any type of fencing is either a physical barrier, or a pain barrier. To provide the best protection against predators, a poultry fence should feature a combination of both.

Let's talk for a moment again about avian predators. All the fencing in the world won't keep a hawk or eagle from swooping down and dining on chicken if the opportunity presents itself. To be fair, most avian predators won't bother chickens. We actually encourage our raptor population here, despite the fact that our chickens roam the property. Since we still have a rodent population, we encourage the raptors so that they'll dine on the rodents. In 14 years, we haven't lost a single chicken to raptors. However, we know that some farms definitely have lost poultry to raptors, so it's prudent to at least plan for that possibility.

There are two basic forms of protection from avian predation – a solid roof, and bird netting. The solid roof allows the poultry to take cover whenever they see an avian predator on the wing. Some poultry that live their entire lives under solid roofing, will never develop the habit of watching the skies. But watch a batch of poultry who have been living on pasture awhile, and the observant flock owner will see one or two of them checking the sky every few moments. As soon as the see the silhouette of a large bird, they'll dive for cover under some solid object. Poultry owners can capitalize on this behavior by providing multiple roofs or simple overhead objects which the birds can hide under. Even something as simple as lawn and patio furniture, picnic tables and other normal human objects out in the yard, can be used for this purpose.

Bird netting is a lot more involved, but it also provides better protection over a wider area. Bird netting can be super-thin nylon netting, or heavier cord, depending on how large an area it's going to cover. Similarly, it can be lightweight enough to need almost no support, or heavy enough to need a rigorous framework. The thickness and thus weight of the material, plus the area spanned by the netting, will determine what type of support, if any, it needs. Super-lightweight bird netting, such as that intended to protect trees from bird damage, works well for small areas. Much larger than a 100 square feet, however, will probably require some kind of frame to keep the material from draping down so low that the poultry will get caught in it themselves. To see what types of frameworks are available, or recommended, check out small batting cages which are available in sporting goods stores and even online. The batting cages use a material similar to bird netting, to keep the balls from leaving the batting cage. That material, and the framework required to keep it supported, would work for a small poultry yard as well.

One big caution with overhead netting or roofing of any kind, is that snow can and will accumulate on most forms of netting. Snow might seem light and fluffy, but it actually has considerable weight. A framework that will very easily suspend lightweight netting in spring, summer and fall, will collapse under the weight of even a small amount of snow or ice. If a flock owner is in a part of the country which can get either occasional snow accumulations or ice storms, consider using that netting as a three-season protection only. The birds won't be out in inclement weather anyway. Consider taking down the netting for the winter, and using smaller structures such as tables and the like, to provide overhead shelter during those winter days when the birds are outside.

For a description of how different types of fencing can be combined for a nearly impenetrable boundary between poultry and the rest of the world, please see my description of the best chicken coop I've ever seen (and the nicely fenced yard), under the Vermin Control section.

Chicks require some additional protection from predators. Granted that many micro-scale and small-scale flock owners will never raise their own chicks. However, day-old chicks are commonly available for purchase from late winter through early summer, with many first-time poultry owners getting started with day old chicks. So a few words about their special needs are probably warranted. Chicks are normally kept in smaller containers or chambers known as brooders, while they grow and reach at least adolescent size. While growing, they are prey to just about every form of predator ever invented. They will even be attacked by other grown poultry, if they have the least bit of chill or become incapacitated for whatever reason. Thus, it is essential for poultry owners to keep chicks and young birds away from everything, including older birds.

Fortunately, a number of living arrangements are possible. First is the question of whether the chicks are being raised by their mother or artificially. If the former, the mother hen won't allow other birds near them, regardless of how they are being raised. If the hen has plenty of room and only a few chicks, she can normally herd them around and keep everyone out of trouble. But if she has more than about four chicks, she can lose track of where they are, or they can lose track of where she is, despite calling to each other. Opportunistic cats, raptors and even rats can use that moment to pounce on the young chick. Alternately, if a hen and her brood are being kept in very small quarters, other birds and predators still need to be kept safely away because she can accidentally trample the chicks if she starts to feel threatened. A 2' x 3' rabbit hutch, dog kennel or water tank is a nice size to provide the mother and chicks with at least a little moving around space, without allowing the chicks to get too far away. If using a dog kennel, the chicks will often move to the darker, rear of the kennel if other animals approach the entrance, thus allowing the hen to stand and defend them without getting trampled in the process. That living arrangement can work nicely as long as the kennel is cleaned frequently, and soiled bedding replaced with fresh. A rabbit hutch can work extremely well, because a wire mesh floor allows the birds' waste to simply pass right through while the birds stand on the wire. This approach will keep them nice and clean, but make sure the mesh is small enough that the chicks' legs don't pass right through it. Also be careful when using this approach for meat birds, because their fast growth and relatively heavy weights on the wire mesh can result in sores on the bottoms of their feet.

For chicks being raised artificially, namely with a heat source such as a lamp rather than their mother, they absolutely positively need to be protected from having other animals get into their brooder. Even cats that normally don't bother birds will look at a fuzzy little chick like a little meal, and the chicks have no defense. Rats won't generally go after healthy chicks but anything that isn't moving, such as a chilled or sleeping chick, is fair game. In this instance, dog kennels don't generally work as housing because there's no safe way to put a heat lamp in such a small space. Rabbit hutches and water tanks are still both very good ideas because the heat lamp can be suspended over the cage or tank at whatever height provides the right amount of heat at floor level. A rabbit hutch with ½" by 1" wire is good protection from other birds and cats, but mice and small rats can still come and go through the wire. Using ½" by ½" wire would keep even mice from going through. A steel or heavy plastic water tank with a lid would also provide decent protection. Chicks could be kept in either type of housing from day-old status all the way through until they are adults, as long as they have sufficient room, and as long as the heat source can be moved further and further away. Raising chicks is specialized enough that it warrants much more general care information than I can provide here. I did however at least want to address the housing needs of chicks, and some reasonably convenient ways to provide for those needs. Folks interested in this approach should check into further sources of information before bringing day-old chicks home.

## *Disease Control*

Poultry are generally quite robust and healthy. Yet like any living thing, they can and do get sick. While most of those illnesses are specific to birds, some avian illnesses can infect human beings. So let's take a closer look at how or why birds get sick, and how to minimize those instances. In particular, let's see how coop conditions can affect the overall flock's health, how disease can come into the coop, how it spreads, and how coop design/maintenance can minimize this problem.

Diseases amongst poultry generally stem from one or more of these sources:

- A buildup of manures

- Broken eggs which then spoil

- Spilled/spoiled feed and water

- Rodents, wild birds and other animals which carry disease into the coop

- Poor ventilation and/or design such that the coop is too hot, too damp, or too stuffy

Let's consider each source, and how to avoid it.

## *Manure Buildup*

No matter what else chickens do, whether they're egg layers or meat birds or just pretty animated ornaments for the lawn, they poop. And they poop a lot. Any coop design must have ways to either allow that manure to pass out of the coop (for instance, with a wire floor), be collected (for instance, under the roosts), be buffered (for instance, by bedding), and most importantly, be cleaned out. If manure is allowed to build up within the coop, air quality, egg quality and chicken health will all be affected. We'll talk in later sections about how various coop designs can affect, minimize and/or control manure buildup, so I won't repeat that information here. Suffice to say that a flock owner will need to decide, preferably in advance, how he or she is going to address manure buildup. If the coop is kept clean, the birds will enjoy fresh air and clean living conditions such that diseases from manure will never occur. However, if birds are forced to live in their own filth, they can and will get sick, repeatedly. For any given coop design, consider how the flooring will either allow manure to pass through to the ground beneath, or contain manure to be removed later. A later section on the use of bedding within the coop will also provide some information on how manure can be collected for use in gardens as a valuable soil amendment, while keeping the coop fresh and clean in the process. Mobile chicken coops and so-called chicken tractors, which we'll talk about separately, avoid the problem altogether by moving the coop so frequently that buildup doesn't occur. Each of these methods has various pros and cons, but they all boil down to either moving the manure away from the coop, or the coop away from the manure. In either case, the birds should never be allowed to live in a buildup.

## *Broken Eggs*

This might sound like an odd source of disease, but it can happen fairly rapidly if not caught early. Eggs are a near-perfect food source, not only for us but for a large host of insects and vermin. When an egg breaks, that leakage doesn't just sit there innocently. Every tiny little creature in the neighborhood that can smell it or sense it will try to partake of its yummy goodness. And considering that insects and vermin don't generally wipe their little feet or wash their little hands before eating, they will track in and walk through that spilled egg in the process of dining on it. Then they'll walk all over the nest, and probably around the coop, on their way to their next meal. A single little set of footprints from some creature after an egg meal probably isn't going to matter much. But hundreds or thousands of them, often enough, is going to spread that contamination all over the coop. At some point, at least one of those little beasties will have been dining on something that was already infected with a germ of some kind. Germs love eggs too, for many of the same reasons we do. And when the vermin introduce the germs to the broken eggs, the germs feast. When the vermin track that egg waste all over the coop, they bring the germs with them. And contagion spreads.

To prevent this scenario, eggs simply need to be collected each day, and nest boxes with broken eggs cleaned out promptly. It's really that simple. For a well-designed coop with easily-checked and easily cleaned nest boxes, this is a very simple task. However, if the nest boxes are hard to get to, hard to look into, and/or hard to clean out, a single broken egg might become a big hassle to take care of. Since most folks don't generally enjoy hassles, that task might be left until "later". And then the above scenario begins to play out. While nest box design, access and hygiene might seem so basic as to be obvious, I have seen coops where the nest boxes couldn't be removed at all, making any sort of cleaning operation nearly impossible without a high-pressure sprayer. And then the whole coop would have been wet for hours afterwards. Or a coop where the nest boxes were on the ground, such that the flock owner only collected them every few days. Furthermore, that particular nest box arrangement was so inconvenient, that only the eggs towards the front could be reached, and it would take a flashlight to see all the way into the back of the nest box (while kneeling down in the poopy bedding). No big surprise that egg collection was often put off until "later", and broken eggs invariably built up in the back where they couldn't be reached. A simple design change could have avoided both of those scenarios. Keep the nest boxes at a convenient height, make sure they can be easily removed and easily cleaned, and change out the bedding frequently. Those three simple steps will go a long way towards keeping broken eggs from occurring, and disease from spreading.

## Spilled/Spoiled Feed and Water

Chickens are not careful eaters – they will gleefully scratch food all over the coop as they eat. It goes back to their ancestral roots as jungle floor residents, where they scratched through the leafy litter to find tasty bugs underneath. Anyone watching a flock of birds moving through a meadow or field, will see this behavior. And any flock owner can see it in the coop too, if there is bedding on the ground. Unfortunately for flock owners, this behavior can also result in a considerable amount of spilled feed if that feed is stored or presented anywhere near the birds' feet.

We talked a little about how birds will try to rake feed out of feeders if the feeders are at or near ground level. So one very successful way to minimize spillage is to keep all the feeders at least a little bit off the ground. Even two inches off the ground is enough to discourage most birds from scratching through it. Minimizing spilled feed will thus also minimize that feed getting wet, and getting moldy. It's the moldy part that we're interested in here. Moldy feed is definitely a threat to birds, and they absolutely will eat feeds which are contaminated with mold spores. Contrary to a variety of old tales, birds have no sense of smell, so they can't smell if the feed is musty or rotten. They'll simply taste it. If the feed is moldy enough, the taste may be off enough that they won't eat it. But oftentimes they will continue to eat until they've eaten enough to make them sick.

Even when chickens can't spill their food, feed can come to the flock already spoiled. The most common reason for spoiled feed is that it got wet at some point – whether that occurred before or after being bagged up. If a flock owner ever opens a bag of feed and it smells off, smells musty, or particularly if it feels hot to the touch, STOP. That feed is already spoiling and should be returned to the feed store immediately. A reputable feed store will exchange it for a new bag of feed. If that happens repeatedly, either talk to the feed store owner or switch to a different brand, or a different store. Feed should always smell fresh, without any sort of dusty or musty or moldy smells or appearance. Anything less is a defective product and should be considered as such. If the bag was Ok when first opened, but water got into the bag or into the bin holding it, and it has started to spoil, STOP. Feeding spoiled feed is a waste of time – the spoilage is robbing the feed of any nutrients, and there's a good chance the birds will get sick. Either compost or throw out that spoiled feed and replace with fresh. Finally, sometimes large bin feeders will get small pockets of buildup in corners and under metal edges, where the feed gets stuck and starts to spoil. If the bin is continually refilled without ever allowing it to fully empty, this buildup won't be visible. Periodically throughout the year, let the birds eat the bin down until it's empty and then check it to ensure there isn't some wad of spoiled feed building up somewhere.

While spoiled feed is typically the source of problems, contaminated water can also be a culprit. If a waterer is located low enough that birds can either kick up bedding into it, or poop directly into it, they will. At which point that water will need to be changed. One very simply solution for this is to keep the waterers high enough that the birds can't kick debris up into the pan. Another solution is to use nipple waterers where there's never any open water to begin with. If using larger tanks with nipples, or larger reservoirs, also check for the buildup of algae and/or mosquito larvae in the reservoirs. Cover any large tank with a light-proof covering to prevent both problems. Water that has become soiled can definitely carry and concentrate disease, and it's a problem that is exceedingly easy to avoid with just a little bit of forethought and/or maintenance. If nothing else, clean the waterer once a week to ensure that there is no buildup.

## *Rodents, Wild Birds and Other Visitors*

Domestic birds purchased from reputable poultry houses very rarely ever bring disease with them into a coop. Usually, if they are carrying diseases they contracted those diseases after leaving the hatchery. If their living conditions have always been kept clean as described above, they can still be exposed to contamination by wild birds, wild rodents and other wildlife. This is where some knowledge of local and regional disease issues can be important. This is also where flock owners will have to decide where to draw the line between "good enough", and fretting so much about unlikely scenarios, that they never get any sleep at night.

Wild birds are known to carry a variety of diseases which can affect other bird populations, as well as human populations. Many of the recent so-called bird flu epidemics may have originally started from poor hygienic practices somewhere, but wild birds picked up that contamination and spread it to other poultry populations. As wild birds migrate north and south during the course of the year, they can pick up and spread disease organisms far and wide. Many times any diseases spread in this manner won't be of great significance. But when there is an outbreak of illness which can impact human health, such as avian flu, or even when it's simply a disease particularly dangerous to birds, wild birds become an important vector in the spread of that disease.

A few years ago, public health agency concerns about this risk led some county and state health agencies to consider "liquidation" of flocks as one possible strategy to control future outbreaks. What that means is that if a certain avian disease broke out in a certain location, and that disease posed a threat to human health, that state agency would be authorized to find and destroy all nearby wild and domestic bird populations, to ensure it didn't spread further. This "scorched earth" policy was and remains controversial, but flock owners need to be aware that it is still a strategy that may be used in the future if a serious disease outbreak occurs.

Even when a disease outbreak is not a threat to human health, commercial poultry operations can request (and sometimes achieve) the liquidation of nearby hobby flocks to stop the spread of avian disease. This approach is intended to protect their multi-million dollar egg farms or meat bird farms from contamination. While such maneuvers almost always create a great deal of controversy nearby, flock owners need to be aware that nearby commercial operations won't smile warmly upon saving the backyard flock, if their commercial operation is threatened.

So now that we've established that diseases can go from wild birds to domestic birds, what to do? We find ways to keep wild birds out of our domestic flocks. The specific concern in this context is that wild birds will be attracted to the poultry feed. A free meal is a free meal even if you're just a sparrow, and that sparrow enjoys a free meal as much as anyone else. When poultry feeders are left out in the open, they will soon become neighborhood wild bird feeders too. Keep those feeders in the coop, and/or provide some kind of protected area such that only the poultry can use them. Either method will go a long way towards keeping out wild birds. Also keep in mind that wild birds are also more inclined to fly into a building through a window, rather than walking in a door. Since poultry are ground-based birds, keep any "bird door" near the ground. If the coop has both chicken doors and people doors, it's OK to leave the chicken door open while keeping the people door closed. That single strategy will usually prevent wild birds from flying into the coop to feast on the feed.

In addition to wild birds, rodents and other wild animals can carry and spread diseases. Raccoons, skunks and possums in particular are vectors for a number of diseases that can affect both poultry and human health. Keeping them out of the chicken coop and chicken yard is thus a good strategy for disease control. We've already talked about ways to keep them from gaining access to the bird coop and bird yards, in the Predator Control and Vermin Control sections. I won't repeat that information here. But everything written in those sections applies here as well.

One final note on this topic. Sometimes the most dangerous visitor to the chicken coop, is the well-intentioned human visitor. After all the other steps we've taken to prevent disease, we can't overlook the risk to our birds from off-property human visitors. Consider for a moment how many places people walk around – they walk on city streets and sidewalks. They walk through rain puddles and parking lots. They walk through yards with cats and dogs and wildlife galore. They walk through other people's farms, and fields and valleys and mountains and caves and 1001 other locations. And worst of all, the flock owner has no way of knowing if any of those places, left some germs on their shoes. This is the heart of a new concern for what has been called bio-security: the prevention of contamination from human sources.

One of my livestock mentors is very fond of telling the following story. He has a small hog farm not too far away, and he gets his share of county and state government officials. Being a stickler for appointments, he had a standing request that anyone from those agencies make an appointment with him prior to their visit. One state agriculture department official didn't honor that request, and showed up at his farm gate unannounced. In the time it took for my mentor to realize he had a visitor, the official had gotten out of the truck, opened the farm gate, driven through, had closed the gate, and then had stepped out of the truck again, all while wearing boots that were covered in mud from someone else's farm. My farm mentor was not amused. He very matter-of-factly told the official to remove the boots. The official asked why? My mentor then went to the nearby garage, got a small propane torch, lit it, and told the official very matter-of-factly that those boots were a bio-security hazard to the health of his pigs, and as such they needed to be destroyed. He assumed the official would not want to be wearing the boots when he destroyed them with the torch, and was merely trying to protect the official from being burned in the process. The official hesitated a moment, not sure if this was a joke, but then hastily vacated the boots, which my mentor then torched on the spot. That official was in turn not amused at the time. They concluded their business and the official left, minus the boots. Upon further reflection, that official admitted that yes, my mentor was correct not only in recognizing that threat, but in dealing with that threat in that manner. Not only did that official make appointments from then on, but that state agency also began a policy shortly thereafter of using disposable footwear for farm visits. The new policy required that any official visiting any farm, must use a new, disposable set of footwear, and change footwear between visits to subsequent farms. Welcome to the world of biosecurity. It's a hassle, but the threat is real and the stakes can be high.

We've already talked about how wild animals can spread disease, but so can people. Thankfully, we don't need to torch visitors' footwear as they stand barefoot in the driveway. Many farms are now requiring that visitors either take off their shoes in exchange for footwear provided by the farm owner, or wear disposable plastic booties while walking around the farm. This might seem over-the-top silly, but multiple tests have repeatedly found that folks track a heckuva lot more garbage on the soles of their shoes than anyone wants to acknowledge. Simply preventing those soles from touching the flock owner's property, will go a long way towards ensuring that germs from other locations don't come into his or her flock yard. Other farms and flock or livestock owners go another step, and insist that anyone with cold or flu-like symptoms simply don't visit the farm. That would be common courtesy anyway, but sometimes folks don't realize that their minor cold can be spread to a farm's working animal population. And some farms, ours included, deliberately minimize visitors of any kind, or restrict visitors to alleys and walkways where they are never in direct or even nearby contact with the animals. That might seem harsh, but the risks of contamination from careless human indiscretions are simply too high.

"That's absurd!" would be the very common, and understandable, response from many potential visitors. The evidence is clear, that visiting human beings pose serious threats to the health of working animal populations. Our livestock create our income, and as such we protect their safety and security as someone else would lock up their storefront or desk. That is our choice and our preference. However, a flock owner has to decide for themselves how comfortable they are with the risks of contamination, versus the joy they would provide by allowing friends, neighbors or family members to come visit their flock. We understand that dilemma, and we have at times allowed friends and family members to come visit the farm specifically to see the livestock. We try to ensure that the visitors don't make bodily contact with the animals, and we have boots for the nieces so that they can walk safely around (per our preference), without getting muddy (per their mother's preference). So there are ways to compromise. I bring this up here not because I think every flock owner should lock away their flocks. Rather, I want folks to understand where risks come from, and how they can be either avoided or reduced. Ultimately, the flock owner must decide how to manage their own animals.

## *Poor Ventilation*

This might seem an odd form of disease control, but let's consider a few things. First, we know that birds have a fairly wide comfort zone, but even they can get too hot or too cold. And any animal that is consistently too hot or too cold, will be stressed enough that they will eventually get ill. So our coop design should provide for either a nice breeze on hot days, or protection from cold, wet conditions, or both as the conditions change. Many coops that have a nice open floor plan with a few windows and/or one open side, will have no problems with either of those situations. However, I have seen coops which are essentially closets with roosts and nest boxes. No chance for air exchange, so the coops got infernally hot in summertime and the air was cold and clammy in winter. Both of those situations are ripe for disease outbreak. In that particular situation, the birds were typically outside all summer long, so they weren't so prone to the heat buildup. Yet when winter came around, the birds would repeatedly get sick. That is a coop design problem.

A very closely related problem is the buildup of ammonia, from accumulated wastes. Chicken manure is very high in nitrogen, because chickens don't urinate separately like most animals. The urine is a solid white paste which is excreted right along with their manure. When that is exposed to the atmosphere, it releases ammonia. If the air inside a coop is stale and no fresh air can get in, that ammonia can build up over time until it is toxic.

While home flocks are generally outside for most of each day, inclement weather will drive them inside frequently in winter. That's when most health issues occur. The birds might already be a little chilled from the colder temps. If they are also subject to either cold damp conditions inside, and/or a buildup of ammonia in the coop due to poor ventilation, they will get sick. So a coop design which features good air exchange, throughout the year, is a must for healthy birds. As a flock owner considers various coop designs, or designs a custom coop, generous air exchange should be near the top of the priorities list. That single design feature will go a long ways to ensuring flock health.

## *Heating and Lighting*

Happily, most birds most of the time require very little supplemental heat. When chicks are growing up they definitely need supplemental heat, because they can't immediately generate their own. That heat is typically supplied either from snuggling close to Mom or from an artificial light source such as a brooder lamp. Once birds are old enough to have true feathers, they need less and less supplemental heat until finally they need none at all. Adult birds which are healthy, dry and out of the wind, do not need supplemental heat except possibly during sustained, below-zero conditions.

That being said, poultry enjoy heat on cold winter days, and they will bask in the sun year-round until temps are well up into the 80's. For southern flocks, supplemental heat is almost unheard of in the coop. For northern flocks, many flocks get by just fine with a shelter that allows them to get out of the wind, rain and snow. If the flock owner really wants to spoil his or her birds, he or she could set up a small heat lamp to shine on the floor of the chicken coop during really cold days. The birds will gleefully bask underneath it. The flock owner must of course take precautions that the heat lamp is installed such that it cannot be landed on, scratched against, pecked at or otherwise pulled down, to become a fire hazard. Also for northern flock owners, be aware that comb size/shape can make a big difference for a bird's comfort during the cold. If a bird's come is narrow and tall, that shape is very prone to frostbite during temps that go much below 28F. Roosters in particular will try to tuck their entire comb under a wing whenever it gets cold, but sometimes the comb is simply too big to fit. As a result, the upper tips of the comb may turn black and fall off. This is undoubtedly uncomfortable for the short-term, but it's not generally a serious problem. Keep an eye on birds with tall, thin combs during cold weather, and if they start doing a lot of head-shaking or tuck their heads under their wings even during the day, their combs are starting to get too cold. At that point, a heat lamp may make the difference between a comfortable bird and a frostbitten comb.

One alternative to this problem is to select birds which don't have regular combs. Many poultry varieties have smaller combs which are not nearly so prone to frostbite. They may look odd, but beauty is as beauty does, and a frost-free comb in the dead of winter is a nice benefit. A bird in pain won't lay eggs, so if seriously cold weather is a realistic concern, small-combed birds may produce better through the winter.

On the other end of the spectrum, high temperatures can have serious health effects for birds. Poultry are able to insulate themselves extremely well against the cold thanks to their feathers, but they can't shed heat very well. As a result, temperatures much higher than 85F become a problem for birds. The smaller, lighter birds seem to have less overall problems with heat, while the larger, stockier birds seem to suffer more on hot days. A flock owner in hot climates can help at least reduce this risk, by keeping the air moving through the coop. Birds will seek shade in hot weather and if the coop is the only shade, that's where they'll go. If the coop is even hotter than the outside, the birds might have a tough choice – a hot stuffy enclosed space versus direct sunlight. Keep the air moving through that coop, or better yet design it such that air moves through naturally. That way the birds can seek shade and benefit from the moving air, which will carry at least some of their body heat away. Other options are to supply additional shelters in the yard, which provide shade but are open on all sides to completely eliminate any heat buildup. One very nice idea, if a flock owner has this option, is to provide a retreat into a cool basement or small root cellar. The earth itself prevents that space from overheating, and flocks would luxuriate in the moderate temps while it's baking outside. A third option is to provide a mister in either the yard or the coop. The mister would cool down the immediate area through evaporation. It would not feel as good on the chicken's body as a fine mist feels to us, since the feathers block the mist from actually contacting the bird's skin. But well-designed misting systems can drop the ambient temperature 5-10F or more. Most small flock owners won't go to such lengths and the birds just tough it out as best they can. However, keep in mind that high temps are enough of a problem, that commercial hatcheries and laying farms have complex equipment to keep the birds at a nice even moderate temperature. If the small flock owner provides even a few shelters or other cool retreats, that can help the home flock stay on regular egg production through the worst of the summer heat.

Lighting for the home flock is a lot easier to provide, but just as important. As we've already discussed, poultry need at least 14 hours of daylight to trigger the hormones which control egg production. The typical solution for that is to provide supplemental lighting in the coop. Some flock owners have lights going all day long. This does not make the birds lay more, but it does make midnight checks a lot easier to accomplish. Other flock owners have their lights on timers, such that they provide supplemental lights for a few hours in the morning and evening. This has the double advantage of not only ensuring the birds continue to lay, but it also makes morning/evening chores a lot easier. The flock doesn't need a lot of additional light; if the coop has enough light such that the flock owner could read a book inside the coop, that's enough light to trigger the egg-laying hormones. A reminder that if the hours of daylight drop below that 14 hour amount, not only will the birds stop laying, they will also start molting. Many flock owners weave this biology into their yearly flock management cycle, in the following way. During summer of any given year, the flock owner buys in the next batch of day old chicks. Summer is a perfect time to raise chicks because the weather is already nice and warm, so it's easier to keep the chicks warm as well. The chicks will grow and reach egg-laying age right around the first of the year. If there has been no supplemental light, they won't lay, or they'll lay poorly. But as soon as that supplemental light is provided, they start laying like there's no tomorrow. Then they lay consistently that entire year. As they reach the end of that year, they are coming up on 18 months of age. If the flock owner has repeated the cycle and brought in another year's worth of new chicks, then he or she can end the supplemental light around the first of December, and the mature flock will almost instantly stop laying, and go into molt, as the next generation of young birds continues to grow. The flock owner can then either sell the mature birds, or let them go through their molt, and return the supplemental light 60 days later. By then, his or her entire flock will have molted and be ready for another laying year, and the new birds will be ready to lay as well. Instead of molt occurring unpredictably throughout the year, and those birds going out of production in the process, the flock owner can limit molt to that specific window of time.

Small flock owners may not care so much about laying and molting predictability, but that technique is there if folks want to use it. If the flock owner simply wants eggs most of the year, a very easy schedule would be to provide supplemental lighting from February 1$^{st}$ through April 15$^{th}$, then again from August 20$^{th}$ or so through until December 1$^{st}$. Let the hens rest during December and January, then fire up the lights again in February and everyone can get back to work. I should repeat here that once a hen molts, she will never again go back to the level of egg laying production she had prior. Yet many hens, given a comfortable stress-free place to live and a good diet, will continue to lay for years. It's ultimately up to the flock owner to decide when a hen's career is over.

## *Supplemental Heat*

We have already touched a bit on both of these topics, but let's give them each a little more details. In terms of heat, poultry are fairly hearty creatures, with a good supply of insulation via their feathering. Most areas of the country won't ever get cold enough to require that the birds receive supplemental heat. Poultry in good health can easily go down to 0F on occasion with no supplemental heat, as long as they have a shelter from the wind and precipitation. In that still air, their feathering works wonders to protect them from the cold. However, if they are exposed to the wind, for instance out on pasture with only a roof and no windbreaks, they will huddle as best they can, packed closely together on roosts or huddled on the ground. Being on roosts in that situation is better than being on the ground, because contact with the ground would further rob them of their body heat. That type of situation is definitely to be avoided because that makes them more prone to any sort of illness germ that happens to wander by. Give them a roof, a roost and a windbreak (for instance, have the roost surrounded by at least two walls, preferably three against prevailing winds) and they can tolerate cold spells much, much better.

Sustained cold can become a problem not for the birds' overall health, per se, but because their combs can freeze. We've seen comb damage here when temps were well below freezing, for days at a time. Our largest roosters have been the most prone, with their large combs. The comb tips would turn black and eventually fall off, despite the birds trying to protect the combs under their wings. That type of weather is rare for us, yet it happens say once every five years or so. Given that, our coop at the time didn't provide quite enough heat to protect the birds' combs, yet our next coop will provide some heating options for those uncommon events.

Heating options in those events include several low-cost choices, including ceramic radiant heating elements that can be mounted on a ceiling out of pecking reach, and/or more common heat lamps (either red or white) which shine a beam of intense light down into a small area. Reptile supply companies have a wide range of the former, typically used in terrariums for large snakes and lizards, while hardware stores often have the latter. Feed stores also often have the heat lamps because they are so commonly used as brooder heating sources, for both chicks and piglets. The lamps have the added benefit of providing light which we'll touch on next, yet have the same risk of breakage as any other type of bulb. The ceramic heating elements are more expensive, but much, much more durable.

Standalone heaters, such as the indoor-rated kerosene heaters and electric so-called "milk room heaters", are not recommended for birds for several reasons. First, they have open elements or flames, which can ignite anything that comes in contact with them. Even if they are well shielded from actual contact with the birds, any kicked-up bedding and/or loose feather wafting through the air can get into the heater and ignite. Secondly, birds will want to roost on top of them because that hot air will feel magnificent under a bird's scaly feet. But if the birds defecate on the unit, the electric unit can blow a circuit, or the kerosene heater can either become fouled or simply stink to high heaven as the manure bakes on the hot heater housing. There is also a third risk of having the heater being knocked over as the birds crowd around it. All told, these two options are simply not good ideas for the chicken coop or yard.

Some folks will take a slightly different route, and house the birds in a heated room or greenhouse, where the entire space is heated. That might be a luxury for the birds, but it can work well if that space is used for other purposes, or if it is designed to be super-efficient. For instance, a greenhouse or hoop house can heat enough during the day, and retain enough warmth through the night, to provide all the heat a flock may need in many parts of the country. The hoop house or greenhouse covering also provides a near-ideal shelter from the winds, IF the house ends are also closed off. However, if the ends are left open, cold winter air can blast through that structure like a wind tunnel. So definitely close off at least one end of such houses if they are being used to house poultry in winter.

Radiant flooring in stud construction houses may be extremely efficient, if that system is already in use elsewhere on the property, and if the coop doesn't have a lot of bedding on the floor. That approach would not only keep the coop warm throughout the structure without risk of fire, but it would also keep the waterers ice-free. The only time I've ever heard of this approach being used, is when a farm had their entire barn set up with radiant floor heating. All the animals thoroughly enjoyed the heat, and it was a very cost-effective way to provide that heat through a winter which was typically fairly harsh. For an adventurous family that wanted to experiment with this approach on a small scale, there are enough DIY kits available in home improvement stores now that a system like this could be rigged up. However, it is definitely not a common approach, so there are not many models to follow in terms of how to do it right the first time.

I'll toss in several additional uncommon approaches to keeping birds toasty in winter. One would be to house the birds immediately adjacent to a thermal heat sink of some kind, which would moderate the surrounding temperatures. For instance, large tanks of hot water (at least several hundred gallons' worth) could put out enough heat that an entire flock of birds could live immediately adjacent, in perfect comfort. Garden-variety laws of physics say this approach could work quite well, but the design and implementation are more than most people want to mess with. If folks already have large above-ground cisterns or reserve tanks for their household or farm water supply, this could be made to work. But it's probably not cost effective on a small scale. Similarly, some folks who are really into composting, have built animal shelters immediately adjacent to long compost windrows or large bins. Water piped through the windrows or bins will heat up to 150-200F, and have been piped back through the animal shelter as a radiant heat source. A family would then have to maintain the compost pile to provide that constant heat, and that by itself could become its own chore. But if folks are already creating those piles, that's an option. And finally, poultry will naturally seek out other larger creatures if those creatures are warm, such as hogs and cows. There are plenty of farms where the larger animals put off enough heat to provide comfortable environments for poultry. Since this guide is more for smaller homes and properties who are not blessed with larger livestock, I don't see this as a very likely scenario either. I include them here not because they're common or frequently practical, but rather because they did work in some circumstances for some folks. It just goes to show there's a lot of different ways to provide heat, if a person gets creative.

## *Supplemental Light*

Let's talk now about lighting. As we've already mentioned, poultry do need a certain amount of daylight in order to lay. They don't need lights right in the nest boxes (in fact, they prefer the nest boxes to be shaded or downright dark). Rather, their immediate living environment needs to be at moderate daylight intensity, for at least 14 hours per day. The rough rule of thumb is that if a person has enough light to read a book or newspaper, the birds have enough light to lay. So, how best to provide it?

Providing sufficient light will be determined in part, by how much floor space area and building volume needs to be lit. For instance, a small coop of 20'x 20'or smaller, can be sufficiently lighted with a single 60 watt incandescent or equivalent compact fluorescent bulb. A single shop-light fixture, featuring one or two 36" long fluorescent tubes in an all-weather housing, would also be perfectly sufficient. If the coop gets much bigger than 20'x 20', the same fixtures can be used but the building would need more of them. We currently have a 20 x 30' coop and our single 60W fixture is right in the middle, where the birds live. The far ends are definitely darker, but we use those far ends for storage, so those lower light levels aren't a problem. For birds housed in something like a greenhouse, where the structure covering allows light to pass through, it may be helpful to either put reflectors up around the lights so that more light is directed towards the birds, or use a white covering rather than a clear covering. A hoop house with white film is a wonderfully bright environment to work in year-round – no direct sun in summer, and a cheerful bright environment even in the dead of winter. That is my own personal single favorite covering, of all the coverings we've ever used. A covering like that will bounce an incredible amount of light around the building's interior, such that even a single bulb will provide a huge amount of light for both the birds and the birds' keeper. Similarly, a conventional stud construction coop, painted white inside, will have the same effect. It might seem a very minor detail, but it can provide an incredible boost when the sun has disappeared behind clouds for days or weeks at a time, and everything is dark and dreary. It might seem ridiculously indulgent to worry about giving the birds a cheery interior, but remember that a happy chicken will lay more eggs. And a flock owner is feeding them one way or another. So more eggs is generally preferred, in which case a happy flock is also something to at least strive for.

For truly large structures, probably larger than anyone reading this guide will ever need, a high-efficiency lighting system is warranted simply to cut down on the lighting costs. I refer here to buildings which are 30' x 50' or larger, where multiple light fixtures are needed. I've seen a series of shop lights or other fluorescent tube fixtures used in those circumstances, and they are sufficient in terms of providing enough light. Yet high efficiency lamps, such as metal halide grow lights used for the greenhouse industry, provide a lot more light, at a much lower operating cost. If a prospective poultry grower already has access to such lights, for instance if he or she wants to raise a flock of birds in a spare greenhouse, then by all means use what's available.

## *Bedding and Waste*

This particular topic is important for two reasons – first, proper bedding can minimize waste such that it's hardly a concern, while improper bedding (or lack thereof) can turn poultry into a stinky filthy disease-spawning mess. Secondly, soiled poultry bedding can become a wonderful amendment for the garden (or sold as such) if it's properly used, stored and dispensed. Many flock owners also have gardens, where compost-enriched produce is as important a flock product as the eggs. So let's take a little time to consider our options here.

First, a review of poultry manure. Poultry manure is unique amongst conventional livestock species, because the bird's bodies excrete the urine as a solid along with the fecal matter. That makes poultry waste one of the richest sources of nitrogen possible. As such, it's extremely reactive, both chemically and biologically. We've already touched on this topic a bit in the context of disease control, so we won't revisit that aspect of things here. Suffice to say that if poultry manure is allowed to build up, it becomes an eyesore, a stink source and a ripe breeding ground for all sorts of nasties. Yet if that same manure is gathered, stabilized and then either used or sold, it can be a veritable gold mine of plant nutrition goodness. Bedding is the single factor that determines how that scenario will play out.

Bedding is perhaps a misleading word for this topic, because "bedding" typically refers to materials which are used for the individual's comfort during periods of rest. In the context of most forms of livestock management, bedding is used not only to provide for comfortable resting environments, but also to soak up urine and feces. However, birds have no need for bedding in the traditional sense, since they sleep up in the roosts. There may be an argument that they need loose litter on the ground to stir through as an instinctive behavior, and indeed they will do so if given the chance. However, the single most important feature of bedding in chicken coops, is to capture and stabilize the poultry manure in some way, then remove it from where the birds congregate. Even if a flock owner has no garden and no aspirations to ever grow a beautiful head of lettuce, poultry waste needs to be dealt with in a way that won't become an eyesore or a neighborhood nuisance.

Given all that, it would seem that bedding would need to meet some very specific criteria. Indeed it does, but there are three very different branches of thought depending on how that manure is going to be used. One philosophy is to use sand as the bedding material, which has a variety of pro's and con's. The single most distinctive feature of sand bedding, is that it's intended to be put down once and never moved or removed. It may occasionally need to be replenished, but it's not intended to be taken out and replaced on a frequent basis. Folks who put down sand bedding, expect it to still be there 5-10 years later, and their management strategy reflects that. We'll talk about that management strategy in a moment. A second philosophy is to use straw, wood shavings, sawdust, wood pellets, shredded paper or other carbon-rich, absorptive materials. This approach also has pro's and con's, but its main distinction is that it is definitely intended to be temporary. Fresh bedding is brought in and used until a sufficient amount of manure has accumulated. Then the soiled materials are removed and replaced with fresh. Flock owners using this approach must have or be able to acquire a constant supply of new bedding, and a management strategy for removing, storing and then using the soiled bedding. The third option is to not use bedding in any way shape or form, instead allowing the manure to build up in place, for instance over a pasture or garden bed. Then instead of the flock owner removing the bedding, he or she moves the coop instead. Since these are such different approaches, we'll look at them one at a time.

Sand has some interesting advantages as a bedding material for birds. As mentioned above, it's intended to be put into place once, and then generally left in place for the rest of time. Some flock owners then scoop out the waste similar to how a kitty owner would clean a litter box. This still means that manure must be dealt with, but the volume of manure is much, much lower than if it were mixed with bedding. The manure could still be piled up and composted somewhere other than the coop floor. Or it could be applied directly to a garden bed that doesn't currently have plants in it. Fresh poultry manure has very high nitrogen content so it should not be applied to plants directly; it would burn the roots and set them back if it didn't kill them outright. However, since there would be so little volume, an entire coop of 12 birds could be cleaned with a single wheelbarrow every few days, perhaps even once/week. Compare that to removing mixed bedding, which could amount to several wheelbarrow loads per cleaning. Sand also has the interesting feature of being rodent-proof, because rodents can't tunnel in it. We discussed that in some detail in the rodent control section. If rodents are a serious problem, and the flock owner wants to minimize poop scoop duties, sand might be a really good choice. One more note about sand – if used in an exercise yard which is open to the elements, the manure wouldn't even need to be scooped. Sand is a natural biological filter, and can support an amazing amount of beneficial bacterial and fungal colonies. Those colonies would make short work of any deposited manure, but they'd need moisture to do it. For this reason, sand within a coop would still need to be scooped, but sand in an outside yard would not. That can be a major benefit for folks who really don't want to mess with the manure as a soil amendment.

Straw, wood shavings, sawdust, and the like are all classified as high-carbon or carbon-rich materials. That means that each particle within those materials has a tremendous amount of carbon in it. That make sense, since most wood and plant materials are composed of cellulose, which is very carbon-rich itself. Here's where things get interesting for flock owners who are also gardeners. Soils need both carbon and nitrogen, but they need them in specific ratios. Compost is such a nice soil amendment because it has that ideal ratio built right into it. The ideal composting C:N ratio ranges from 20:1 to 35:1. In other words, 20 to 35 bits of carbon-rich material for every bit of nitrogen-rich material. Step back and look at what our bedding does – a lot of bedding is used to soak up a little bit of manure. If a flock owner is even moderately careful about removing soiled bedding and storing it somewhere protected from rain and snow, he or she has a rich compost pile just begging to happen. And the chickens even do the preliminary work for the flock owner, with their constant scratching and stirring. If a flock owner has any gardening or soil-building ambitions whatsoever, a well-bedded coop is one of the very best ways to build a near-ideal compost pile. Even if a flock owner has no such ambitions, that compost can be sold to gardeners who need that material for their own gardens, yet don't have their own birds. We have been approached a number of times by friends and neighbors who want to buy our chicken manure, and this is why they want it. Sorry, folks, but we already know how valuable it is, and we use it all here.

The third general option for manure management is quite different from these first two – namely, removing the coop rather than removing the poo. The single most popular method for this is the so-called chicken tractor, which we'll cover in more detail in a moment. For now, suffice to say that a chicken tractor is simply a small mobile pen, which allows birds to be frequently moved to fresh ground. Any manure deposited is left behind to break down into the soils. The birds are kept clean because they're moved to fresh ground so frequently, and whatever ground they're on benefits from the concentrated but infrequent application of the manure. It can be a really nice way to keep both the birds, and the property, quite happy. We'll talk about chicken tractor designs and sizes later on. A variation on this theme is the small coop which is up on stilts, like a rabbit hutch. Those coops are generally designed simply to provide shelter for egg-laying, roosting, and sometimes eating/drinking. The floors of such coops are often made of hardware cloth or other types of wire mesh, such that manure passes right through them. Some flock owners with such coops move the whole coop from time to time. Others have small bins right under the coops, which can be removed, emptied and then put back. Really clever flock owners have bins on wheels such that they can simply be wheeled away to have the manure dumped out, then rolled back into place. Even something as simple as a child's wagon, or a utility cart, can be used for such purposes.

Any given flock owner may choose any given method, based on his/her goals and access to the various bedding materials (or lack thereof). There's no right or wrong answer here, as long as the birds don't end up standing around in their own manure for long periods of time. And as long as the manure has a chance to break down into something useful, as opposed to either building up, or being washed off into some nearby stream or body of water. Folks who keep ahead of their waste management tasks in the chicken coop and yard, have a nice-smelling, hygienic, attractive coop to proudly show off to friends, neighbors, and perhaps the local public health official who is paid to worry about such things. With a little bit of additional effort, the flock owner can also have rich compost to either use or sell. So this topic, while boring, does merit at least some consideration both during the planning phase, and after the coop has already been built. No sense wasting a valuable resource.

## *Fixed or Mobile?*

As we've started to touch on, sometimes it's preferable for a coop to be mobile, rather than stationary. There really are no right or wrong answers to this particular question, but rather a series of considerations for the flock owner. I'll try to pose some of those considerations, so that a prospective flock owner can ask those questions and evaluate the options in advance, rather than after the fact.

*Q. Will the flock have an exercise yard, and will it move seasonally?*
A. If the birds have an exercise yard in addition to their coop, and the flock owner wants to have that yard in a different location each season or each year, for instance to keep the birds from killing the grass, then a mobile coop would be an excellent choice. If the flock will have a fixed yard, the coop can be fixed as well.

*Q. How big is the flock, and how large will the coop be?*
A. The larger the flock the larger the coop, and thus the more work it will be to move it. This doesn't necessarily rule out the use of a mobile coop, but it does require very careful consideration of strong yet lightweight

materials. Not only will the coop need to resist rain, wind and snow but it will need to resist the forces exerted during each move. One very good option in this instance is either a coop on wooden runners, or a lightweight metal hoop house with reinforced corners. Some commercial growers move relatively large hoop houses (I know of mobile 20' x 30' hoop houses), on tracks. They cycle through growing areas this way, and a flock owner could do the same. Just remember the larger the coop, the more horsepower needed to move it. The flock owner will need a small tractor, an ATV with a hitch, an SUV or truck with a hitch. We have an 8' x 14' shed on runners which can be dragged with our tractor, but I can't move it by myself. Some commercially available coops, particularly non-walk-in types, actually have wheels mounted on one end so that a single person can move them. Or consider a coop made from an existing trailer, which already has the strong frame, wheels and hitch to allow for easy moves. Also see the last question in this list, about issues when the ground freezes.

*Q. Is the coop going to be sized just for the birds, or will people be walking in and out?*
A. As we saw above, the smaller the coop the easier it'll be to move. A coop that will house only a few birds, say 12 or less, can be sized small enough that it will only fit the birds. In that case, it will weigh relatively little, and will be very easy to move. Beware however that if a small hutch-style coop is up on stilts, those stilts can make the move more difficult than if it was on the ground. Sizing the coop and/or the height of the stilts, such that an ATV or pickup truck can back right up to it and carry the coop, will make a move exceedingly easy. In that case, the stilts will need to be installed such that they can be easily removed and reinstalled. Or perhaps simply put the coop on some kind of stand that can be transported as well. Sawhorses are inexpensive and designed to be moved from place to place.

*Q. Is the coop on flat land, or sloping?*
A. Life gets complicated fairly quickly when working on sloping land. Everything wants to slide down to the bottom of the hill, and will do so at the least little provocation. If the coop is going to be on a hill, either keep the coop stationary, or build a very narrow mobile coop and keep the long axis along the contour of the hill. Don't put wheels on it, or someday that

coop will find a way to roll right down the hill. Be extremely cautious when moving it. If the coop is mobile, and oriented along the contour, any moves can then be forward or backward without changing elevation. A long, narrow coop in this instance may be preferable to a square shape, even though the square shape would be cheaper to build. Also be very conscious of keeping the coop level, particularly regarding the slope of the roof. Any mild slant either uphill or downhill will result in much sharper, or much shallower, roof pitch. That can result in more snow load, or heavier rain runoff in a small area. A coop on runners, leveled on the outside via shims or blocks, and level inside with an interior floor, can be a nice compromise. But be aware that small animals can and will take up residence underneath the coop since there will be room under that downhill portion of the floor.

Q. Does the flock owner want access to the manure for compost?
A. Some folks love the idea of gathering that compost, so a stationary coop with a wood, tile or concrete floor would make gathering soiled bedding extremely easy. This is ideal if the flock owner has a dedicated compost pile, and he or she wants to make regular contributions to that pile with the soiled coop bedding. On the other hand, if the flock owner wants only a small flock and has raised beds or wide beds in the garden, a mobile coop can be set up right over any vacant beds. Bed the coop down with sawdust, shavings or the like, let the birds poop right into it and stir it up, then move the coop forward a bit and start the process over. Before long the flock owner will have a freshly fertilized and mulched garden bed, with almost zero effort. No extra compost pile required. Just be aware that a properly set up compost bin can turn waste products into safe, disease-free and weed-free compost in a matter of weeks. A coop built over a garden bed, with the birds pooping into the bedding and then moved, would leave behind nice mulch but it won't be true compost. It will need time to decay down into compost.

Q. Does the flock owner have access to sufficient quantities of either wood chip products, or sand?
A. Bedding cost can be a big motivation to either select, or avoid, one of these options. If a flock owner has ready access to carbon-rich bedding materials, the bedding the birds down with those materials may be extremely cost effective. Particularly if he or she wants the resulting

compost. Similarly, ready access to sand might be extremely attractive because the flock owner will only need to bed the birds down once, and then he or she is done. Particularly if rodents are a serious issue, that sand can avoid a lot of problems from the very start. If both of those materials are hard to find locally, and/or cost an uncomfortable amount, then a mobile coop might be the answer.

*Q. Does the flock owner have a garden, lawn or pasture that needs to be fertilized?*
A. Poultry will range rather far from the coop if given the chance to do so. Ours readily range at least 100' away from our coop during nice sunny days; less on stormy days when it might start raining at any given time. If the flock owner wants the birds to fertilize the front yard, for instance, a fixed coop in back and some fencing to funnel them and contain them out front would give the birds ready access to that lawn area, without daily or weekly coop movements. However, if the flock owner has a larger area to be fertilized, the coop may need to be moved from one station to another, after which the birds will range around it and cover that new area. For folks who want the flexibility of the mobile coop, and don't have the whole area fenced (or want to restrict their birds to one small area at a time) check out some of the electrified bird net fencing which has been commercially available for about 10 years. It works quite well to keep birds where they belong, and does decently well keeping some (but not all) predators out. Premier 1 Supplies carries a very complete line of poultry net fencing for exactly this purpose. Visit www.premier1supplies.com for more information on their products.

*Q. Is the flock owner interested in selling compost?*
A. If the answer to this question is yes, then the flock owner will need to use bedding, and will need to have a cost-effective source for it. Manure by itself can certainly be composted, but the volumes would be so amazingly small as to be hardly worthwhile. Remember that when materials compost, they lose volume; sometimes by as much as ¾ of the starting volume. So manure without bedding would amount to a tiny little volume of compost. A plentiful supply of bedding, on the other hand, will provide a nice sideline product. Be sure to learn and follow composting guidelines so that the finished product is disease-free and weed-seed-free.

*Q. Does the flock owner need to comply with regulations about how bird waste is handled and/or disposed of?*
A Most flock owners will not have specific regulations for how bird waste is handled. However, those folks living within city limits may have certain criteria which must be met, in order to keep birds. If that's the case, and flock owners are well advised to determine that in advance, then he or she needs to either follow those regulations or come up with a Plan B if/when local regulators come a-knocking on the door. This particular issue is often THE issue for local public safety regulators, so it can be a pretty big deal. Playing nice and going along with the regulations can make for a much easier flock owning experience. That being said, some careful research on the options, presented in a logical and compelling way to regulators, can also change local ordinance such that the ordinance reflects scientific fact rather than some lawmaker's trumped-up concerns. If nothing else, know the law. Then a flock owner (or wannabe flock owner) can best decide whether the law should be followed, ignored, or changed.

*Q. How often is the flock owner willing/able to clean the coop?*
A. This might seem an odd question but it can make the difference between a very satisfying flock ownership experience, versus a frustrating one. Some owners are super-diligent about cleaning their coops, and that coop is just as fresh and clean as the day it was first built. I envy those folks, I really do. Other folks have coop cleaning ranked right up there with teeth-pulling in terms of how often they want to mess with it. We're somewhere in between. If a flock owner is one of those who keeps a fastidious home and yard and the car is always spotless, then a fixed coop with either sand or bedding can be maintained relatively easily (and very stylishly). If a flock owner has 1001 other things to do and the lawn regularly goes without being mowed in the summertime, a mobile coop might be the better choice. Whenever the coop floor gets a tad too much, put down bedding. When the bedding builds up high enough, move the coop. Match the coop to what the flock owner needs, AND what the flock owner is most likely to do over the passage of time.

*Q. Will the flock spend time in different areas depending on the season?*
A. This might also seem like an odd question, but it has multiple implications. On our place, after 14 years of running birds and other livestock, we are moving towards a system of summer vs. winter housing.

The summer housing is intended for those months where we won't get snow, we don't generally get high winds, and we don't get a lot of rain. Mud is less of an issue, cold is less of an issue, but grazing, grass control, bug patrol and soil fertility are near the top of our list. Yet in winter, we get lots and LOTS of rain, we get a fair amount of snow, we get lots of wind, and wow do we have mud. As a result, we're moving towards two separate sets of housing for our livestock. During the summer, we are moving towards lightweight mobile coops, out on pasture, with the birds contained by portable electric fencing. For the winter, we're moving towards fixed coops, with a solid floor and bedding inside the coop, and either a covered or at least partially sheltered outside exercise area bedded with sand. This dual nature is not all that unusual, and flock owners may want to consider whether their summertime needs are different enough from winter needs, to warrant different housing.

*Q. Will the flock be following other livestock through a pasture?*
A. This topic doesn't come up very often for suburban flock owners, but it comes up frequently for livestock owners. Particularly livestock owners who either want to add productivity to their pastures, diversify with additional pasture-raised products, and/or provide some natural fly control for their cattle or horses. If poultry will be following livestock to provide any/all of these benefits, then a mobile coop almost becomes a necessity. Those larger animals will graze a larger area during the average summer, than a flock could reasonably patrol from a single central coop. Furthermore, the area under the coop would get too much manure applied, thanks to the roosting birds every night, resulting in dead patches throughout the pasture. Not ideal. This is one of the main situations where chicken tractors came into vogue. Twenty years later it's still a potent idea. Any variety of mobile coop can be made to work in this setting.

*Q. Does the flock owner have access to a small tractor or vehicle with towing capacity?*
A. If the answer is no, then any mobile coop will need to stay either small, and/or lightweight. Happily, a variety of small coops are commercially available, or easily built which feature their own wheels, runners or other methods to make moving a lot easier than it would be otherwise. For any flock less than 12 birds in size, any coop someone comes up with will probably be small enough to move either by hand, or by towing with even

a small car. Just ensure that the coop is on runners or wheels to make it as easy as possible. If a lawn tractor, ATV or small farm tractor is available, the coop can get a lot bigger and still be mobile. One important note, however, that during late winter and early spring, the ground may be so soft that any attempts to move a medium-to-large coop will result in ruts and/or possibly a vehicle that dug itself into the turf. Try to plan ahead such that the coop doesn't need to be moved when the soils are saturated. And note that during deep winter in some parts of the country, even a coop on wheels or runners can freeze right to the ground. More on that in a moment.

*Q. Are storm events and/or winds fierce enough, that the coop needs to be parked in some sheltered area?*
A. if the answer is yes, then the flock owner has two potential issues to consider. First, the coop will need to be as strong as the worst weather conditions require. That might mean super-rigidity for resistance to wind loads, super roof strength for snow loads, or possibly both. Each of those will add weight to the structure, and weight makes it harder to move. Wind loads in particular, since they come from any angle, can really batter an exposed coop blow over an unanchored lightweight coop, or simply blow the roof off. One possible answer to that, particularly with mobile coops, is to park the coop under shelter or alongside windbreaks for the worst possible weather events. Another is to anchor the coop with soil augers and rope, weights, or other forms of tie-down. If weather events can happen at any time (for instance, Nebraska can get tornadoes in summer and heavy blizzards in winter), a strong stationary coop is a good solution. If that coop needs to be mobile during at least part of the year, consider building an anchoring system such that it can hunker down and survive the worst that the weather can throw at it.

*Q. Does the ground freeze in winter?*
A. This event won't make much difference for fixed coops, or for mobile coops that are docked or anchored somewhere for the winter. However, if a flock owner wants to use a mobile coop specifically because he or she doesn't want to deal with changing bedding, be aware that manure can build up inside the coop over the passage of a few weeks, certainly a few months, and cold weather can freeze that coop solidly to the ground. The worse the weather, the worse the problem because the birds will spend

more time inside. Thus manure will build up faster. This problem is worst for heavy, stud-construction coops which are only mobile because they are on runners. Those runners provide a huge area of contact with the ground, all of which must be overcome if the runners have become frozen to the ground surface. If the coop must absolutely move during winter, either mount it or build it on a dedicated trailer, with good ground clearance. Also, move it frequently in winter so that it doesn't have the chance to get bound to the earth after a series of storms or freezing weather.

*Q. Are rodents a big problem, or expected to become a big problem?*
A. We started with mobile coops which didn't have floors, sitting on the ground, providing shelter from rain and snow, but open on the sides for maximum air circulation. Shortly after our birds arrived, the rats arrived. And shortly after the rats arrived, the autumn rains arrived. Shortly after the rains arrived, the mud arrived. Suffice to say we started revising our poultry coop setup. Of all those issues, the rats have been the hardest to deal with. We're a working farm so we always have a lot of animals, and that means a lot of feed potentially available for the rodents. Our first chicken coop, a picture of which can be found later in this manual, offered zero resistance to rodents. They walked right through the fencing, right into the coop and the feeders, and right into the nest boxes. It was pretty miserable. Elsewhere in this manual I've written about how sand bedding, proper fencing, and a hungry hunter kitty or two, can really control rodent populations. Under a heavy infestation, it's very difficult to build a mobile coop that totally resists rodents. That coop would have to be parked on sand, behind some really carefully constructed fencing. The very nature of that setup means that even if the coop could move, the yard can't move with it. So if rodents are a problem, consider using a stationary coop AND sand, AND good fencing, to minimize their access to birds. Or, in situations like ours, use that setup in winter, but then range them in mobile coops during the summer when they'll be out on pastures. We will probably never be entirely rid of them, but that summer/winter setup will help minimize their numbers.

## *Chicken Tractors*

I've set up this special section for chicken tractors, because they are considered a specialty type of chicken coop. Folks may have heard of them without knowing specifically what they are, or where they are best used. So let's talk about them for a moment, before we launch into the next chapter of the manual.

A chicken tractor is simply a very small pen, often less than 100square feet in area, where all the birds' needs are provided in one small compact space. The intent with this design is to move the birds only one pen's width at a time, sometimes once or twice a week and sometimes several times a day. The birds simply walk along inside the pen as it's moved. Most pens have solid frames of either wood or metal, sometimes PVC, with poultry netting or other fixed-type fencing within the frames. A roof typically sits at one end where the birds sleep and eat. Whatever materials, whatever size and shape, that's the general concept right there in a nutshell.

When chicken tractors first came out, they were used primarily for meat birds, which don't generally require roosts. The birds also tend to pack together in close-knit sleeping groups. So interior space is a lot smaller than what layer birds would need. One of the largest issues with meat birds is that if they are not kept on absolutely clean ground, their abdominal feathers get very soiled, very quickly. Their bodies are so large and round that they can't preen those feathers, so that mess just builds up more and more. Thus the search for an easily cleaned pen began. Combine that with a renewed respect for the importance of fresh air, sunlight, living greens and bugs; it became imperative to producers and consumers alike that those birds get outside as early, and as often, as possible. A combination of factors came together and the chicken tractor idea was born.

There's no one right way to build a chicken tractor. Different producers have created chicken tractors from a variety of materials, using a variety of designs, and they all work. The chicken tractor has to provide for each of the issues we've covered so far, in some way – protection for the birds from everything that Ma Nature can throw at them, while allowing them to live as natural birds. With all that variety, several design ideas have come forward as being important regardless of the design. One of them is that they MUST be easy to move. The whole concept of the chicken tractor is frequent movement, and frequent movement doesn't happen if it's a pain to move. Furthermore, the tractor ideally needs to be moved by one person, quickly, without power equipment. The costs associated with driving a tractor, ATV or SUV out to the field twice a day to move a bunch of poultry around, eats up any sort of cost effectiveness from having them outside. The tractors must be able to withstand whatever seasonal weather issues may come up, including both late spring/early winter snows, and ironically heat buildup in summer too. They must be easy to get birds in and out, because they are going to be home to multiple "crops" of meat birds over the course of even a single year. Meat birds are generally butchered at only 6-8 weeks of age, so it's possible that a single tractor will see four or five batches of birds, within a single production season. And they must be strong and durable. Pens are small enough and the birds are visible enough, that most four-legged predators will try to dig under, jump into, gnaw through or otherwise bust up a chicken tractor to get at the goodies inside. That abuse needs a strong frame and durable materials, put together with better than duct tape and zip ties. A well designed chicken tractor should last for several years even with all the movement, all the bird batches, all the weather and predators beating on it.

## *Pre-Made Chicken Coops*

I hesitated to even mention this category of coop, for three reasons. First, amongst all the folks I've ever talked to about needing a coop for some future flock, not one of them expressed any interest in simply buying one. They all wanted to make a coop to meet their particular needs, or to suit their particular goals for function, size, appearance, cost, etc. So a pre-made coop of fixed size, capacity, layout, etc has never seemed to be popular enough to merit much attention. Secondly, of the pre-made coops I have actually looked at, they seemed either gimmicky, or overly heavy, or overly pricey, or all the above. Some of them seemed little more than child's toys. Third, the list of companies offering pre-made chicken coops is fluctuating all the time as some companies stop production for whatever reason, and others begin. For all these reasons, I don't want to provide a list of companies offering these pre-made coops, for fear that such a list would soon be obsolete, overpriced, or simply not what this manual's readership wants.

Yet a complete description of all chicken coop options, requires that I at least mention this possibility. There are also undoubtedly folks who are interested in simply buying something "off the shelf", ready to go. Given this combination, I would encourage folks interested in this option, to search via their favorite internet search engine, on terms such as "chicken coop kits" or "pre-made chicken coops". That will bring up a list of what is currently available near them.

Even if the coop is pre-made, most of the rest of this section still applies – fencing and flooring and predators and environmental concerns are all still issues that must be dealt with. If a reader does want a pre-made coop, I would encourage a careful consideration in advance of the same issues we've talked about in this chapter. How many birds, what sorts of predators, what regulations or neighborhood aesthetics apply, how is the coop going to be cleaned and maintained, etc. The combination of a carefully considered, carefully selected pre-made coop, with a carefully chosen and prepared site, will provide as good a long-term housing answer as any other option listed here.

## *Durability*

This last topic would seem to be obvious, but there's always the temptation to save some money and/or some time when putting these shelters together. Sometimes folks can get away with that, for a time, but inevitably it comes back to haunt them in some way. So I include this comment right after all these other notes, and right before we launch into actual plans, as a warning. If folks skimp on materials, or rush the job, or use a shelter that isn't as rugged as it needs to be, they may save time, money and energy at first but they'll pay it back later. Usually with interest. With our own chicken coops, we deliberately used three-season shelters at first because we wanted to save money. Come that first winter, and I was out there shoveling snow off the roof, every single time we got more than 1" accumulation. When nearby trees would shed small branches, they'd tear through the fabric roof. When strong winds came along, the shelter would jump around on the ground and threaten to flip over unless I anchored it really well. The money and time I wanted to save up front, was spent reinforcing that lightweight shelter. Subsequent shelters were more rugged, but still had issues. Suffice to say we're going back to rebuild our coop yet again, and this time we're going to make it as bomb-proof as we can. I want to spend my winter evenings in my nice warm dry house, rather than outside shoveling off the chicken roof again.

There is one argument in favor of using inexpensive materials, or seasonal shelters. When folks are just getting started, and haven't really learned the ropes yet, it's understandable that they don't want to sink a lot of time, money and effort into building a coop that they might not like (or need) in a few years. We have definitely outgrown several different outbuilding designs over the years, and I'm glad to have had them but won't use them again. There is a balance to be struck between using materials which are probably not going to survive the year, versus limiting the investment for an unproven enterprise. Flock owners will have to debate how best to juggle those two competing issues. A review of the various designs in the next chapter, will hopefully help folks find that magical Goldilocks design of "Just Right". That is my motivation for providing so many different design issues, and plans.

## *Conclusion*

We've covered a lot of ground in this chapter, talking about all the different design features, materials, considerations and options facing flock owners. In some ways this first chapter has been a lot larger than I expected at first, yet ironically I also feel like I've just touched on the surface for a lot of these topics. I encourage folks to continue researching things on their own, beyond what I've written here. There are always new materials, new designs and new options to consider. I'm sure with the passage of time, some new method or material which doesn't exist now, will become the preferred approach. The one thing to remember from all this, is that there are a lot of different ways to succeed with a chicken coop. No one design, no one material, is a magic bullet to solve all issues. And no one approach is hands-down better than others. Each flock owner's setup will determine a lot of the variables and options; those must be the guiding rule when considering coop design. So, with that in mind, let's move forward into the next chapter and start to look at some coop plans.

# SECTION II: CONVENTIONAL COOP PLANS

*A Barred Rock hen high on the roosts during mid-day. Even during the day the birds will often perch on the roosts, as high as they can reach, and survey the action beneath. Photo by author.*

Most farm-sized coops and commercial layer houses share one design assumption – people will be walking in and out of the coop at least once or twice a day, so it needs to be sized for human entry. For small home-sized flocks, that may or may not be necessary. Thus the first design question becomes: will people be going in and out of the coop? If yes, then it needs to be a certain height and a certain width, regardless of length. If no, then the coop can be sized just for the poultry, with a much smaller footprint and cost. We'll consider both situations.

## *Non-Walk-in Coops*

The first and simplest form of chicken coop is that which people don't need to fit into. This type of setup is perfectly acceptable under several different scenarios:

- Up to about 12 birds which have an adjacent, covered outdoor pen

- A "bachelor band" of roosters which don't need nest boxes, with an adjacent outdoor pen

- A brood of chicks being raised up to butcher size, and/or prior to laying

Under these conditions, the space requirements for the birds are so small that it makes sense to make the coop chicken-sized, rather than human-sized, to conserve space and materials. Roosts and a few nest boxes can very efficiently fit into a very small space. There are many different variations on this idea, but the layout generally works out to be a cube or box shape with the roosts on one end and the nest boxes at the other. The box can be made of plywood, stud construction, pallets, or even be a small hoop house with a tarp over the top. It can be on the ground, or raised on legs like a hutch. It can be fastened to a permanent foundation, or mobile. The main criteria with this design, is that feeding and watering either must be via super-efficient layout within the coop, or more commonly is simply done in the adjacent yard. If that's the case, at least that area should be covered to protect the feed from wind, rain and snow. So the materials list will typically include not only the actual coop materials, but additional materials for a feeding station roof as well.

With a non walk-in coop design, the size of the coop itself is so small that the adjacent yard size becomes a consideration. Without getting into larger issues of animal welfare, the vast majority of families who want a family flock also want the flock to be able to get some exercise, fresh air and sunshine. If the coop is sized simply to provide nest boxes and roosts, the yard then has to allow for all those other activities. Furthermore, the yard either needs to be big enough, or the coop needs to be mobile, such that the birds won't turn their yard into a mudpit. This situation is where the non-walk-in type of coop really shines. It is so small that it can be easily moved around from location to location. One very popular option is to put the coop right into the garden over winter, such that the birds can go on bug hunts and stir up the soil between growing seasons. More than one commercial grower keeps a flock for exactly this purpose. They move the flock from one growing area to another, and sometimes station them right inside a greenhouse (or move them from one greenhouse to another), to work the beds, dig up the bugs, and fertilize the ground, all while the beds aren't in use.

One final note on this coop design. If a non walk-in coop is desired for a small laying flock, it is very helpful if the coop is up on a stand, such that the nest boxes are at least 2' above the ground. The birds don't need this height, but the human egg collectors will appreciate not having to get down on their knees to collect eggs. Furthermore, the nest boxes in this instance should have two doors, rather than the more traditional single door. The main door within the coop allows for passage of the hens in and out of the nest box. The outer door can be a simple flap or hatch, which allows for collecting the eggs from outside the coop. This door or hatch should be closed most of the time so that birds don't start to use it from the outside. It should only be opened to collect eggs and/or clean out the nest. The door should also be large enough to enable easy nest-cleaning. We'll look at several variations on this non-walk-in coop theme, in the next two chapters.

## *Walk-in Coops*

The majority of coop designs assume that people will be walking in and out of the coop. This allows for all the normal daily chores – feeding, watering, collecting eggs, checking on bird health, etc. However, it also requires a much larger coop structure for the same number of birds. The trade-off between higher cost and easier chores, is a question each family will have to answer for themselves. Sometimes their neighborhood design covenants or local regulations will dictate some minimal coop sizes. At the minimum, the coop should allow for not only comfortable chicken living, but comfortable human chores as well. We've already covered a lot of those design criteria in the first chapter of this book. If folks have come directly to this chapter looking for plan ideas, but they aren't sure about some of these design criteria, they may want to look through that first chapter and at least skim some of the considerations. It can save a lot of headache later.

The bulk of the plans in this manual will be of the walk-in variety, and sized to allow for reasonably convenient human access. That means at least being able to walk into the shelter (possibly with a low roof, but that particular detail can be altered as needed), and work within the shelter for feeding, egg-gathering, cleaning, etc. Since this is what most folks think of when they think "chicken coop", I've tried to find a wide variety of such plans spanning a lot of different options.

One last set of notes. The plans in this section are courtesy of Louisiana State University's AgCenter website, www.lsuagcenter.com. The plans on that website are available for download in PDF format, free of charge. The plans in this section are used by written permission from LSU's AgCenter, and are intended as illustrations only. For more detail about each set of plans, please visit the AgCenter's Poultry Housing page at http://www.lsuagcenter.com/en/our_offices/departments/Biological_Ag_E ngineering/Features/Extension/Building_Plans/poultry/housing/. If typing or clicking that link is troublesome, simply go to www.lsuagcenter.com and type "Poultry Housing" in the search window near the top central portion of the webpage. That search will bring up individual links to each of the plans listed below.

Most of these plans were developed and/or released by land grant universities, during the period spanning roughly 1940 through 1970. Lumber sizes are fairly conventional to ensure that all specified lumber is commercially available no matter the geographical location. Yet the plans were also assumed to be used primarily for American households and farms. Thus the units of measure are English rather than Metric. No warranty is made or claimed about whether any given design is going to be appropriate for any given location, given the tremendous variations in weather conditions, seismic situations, and even zoning regulations. Each flock owner is responsible for determining if any given design will work in his or her particular circumstances. And finally, none of these plans are stamped by professional engineers, yet most of them are detailed enough that they could be printed out and given to a professional engineer, who could then either issue the stamp for those exact plans, or develop very closely related plans per their current engineering standards. Be aware that this would be done for a fee, but some municipalities require such things. I make that declaration in this section, because many of the plans in the next section are not so specific and would not be ready for an engineering stamp without considerable additional work.

So, all that being said, let's start to look at some of these plans. Just a reminder these are all conventional stud construction plans, sized for flocks anywhere from less than 6 birds, up to 100 birds.

## *Plan #1: Non-Walk-In Coop For 8-16 Birds*

*Plans:*
http://www.lsuagcenter.com/en/our_offices/departments/Biological_Ag_Engineering/Features/Extension/Building_Plans/poultry/housing/Back+yard+Poultry+House.htm

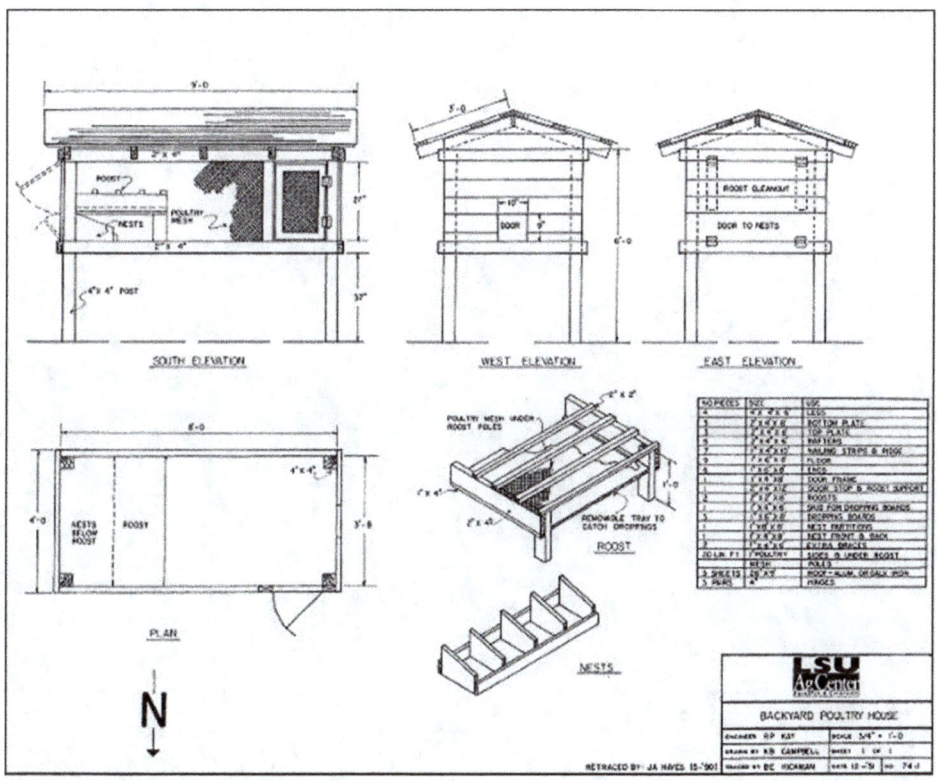

This is a nice little coop which is small enough for convenient positioning, yet roomy enough inside that the birds have room for a feeder and waterer. The one issue is that the row of nest boxes are immediately underneath the roosts. The plans call for a screen under the roosts to catch the overnight manure. But that screen will have to be cleaned regularly to keep it from sagging. If the nest boxes were at the other end of the coop, the screen could be eliminated and manure could pass right through the floor. The small access door to reach eggs in the nest boxes would also need to be relocated. This is a good example of how plans can be used as-is, or tweaked a little to better suit personal preferences. All in all, a very handsome little coop that would work well for many small flocks.

## Plan #2: 20 to 30 Bird Family Flock Coop

*Plans:*
http://www.lsuagcenter.com/en/our_offices/departments/Biological_Ag_Engineering/Features/Extension/Building_Plans/poultry/housing/Family+Size+Laying+House.htm

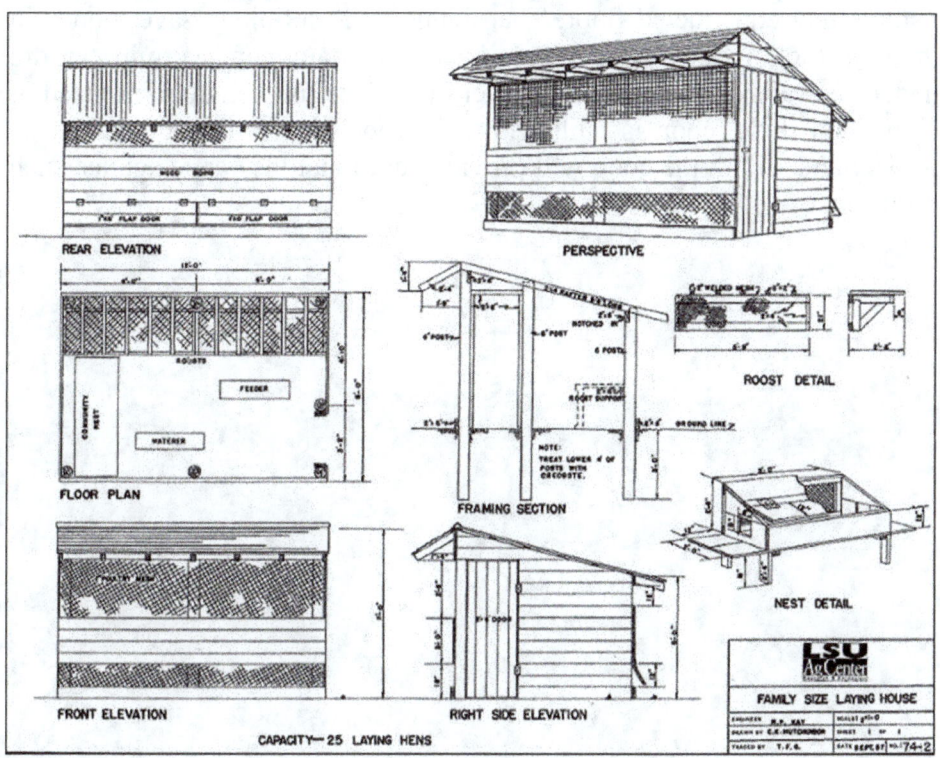

This particular coop can comfortably house up to 30 birds, with ample roosting and nest box space. This larger coop would be appropriate for a family looking to sell some eggs on the side, at farmers' markets or to friends and family, as local ordinances allow.

Two things to notice with this design: the buried posts, and the community nest. This is most definitely NOT a mobile coop. That will increase the labor during construction, but provide excellent stability in stormy weather. The community nest is a very good way to make the best use of limited space. Most individual boxes are sized to minimize the chance for more than one hen to use that space at any given time, but that layout does take more space. A community nest allows more actual floor space for the hens, rather than the space taken up by divider walls. One note of caution however; if one or more of the hens develops the habit of egg eating, the nest will have to be checked and emptied several times a day until the hen or hens forget that particular bad habit. Or those hens can be removed. Another note of caution is that this coop is built right on the ground, without any constructed floor. That feature will definitely save money in the construction. However, a wide host of predators and vermin can dig under the walls and into the coop over time. This risk can be minimized or eliminated if the coop is built with a 3' wide apron of sand around the walls. Sand within the coop will provide additional insurance against such intrusion.

## *Plan #3: 25 to 36 Bird Coop*

*Plans:*
http://www.lsuagcenter.com/en/our_offices/departments/Biological_Ag_Engineering/Features/Extension/Building_Plans/poultry/housing/Poultry+House+10+X+12.htm

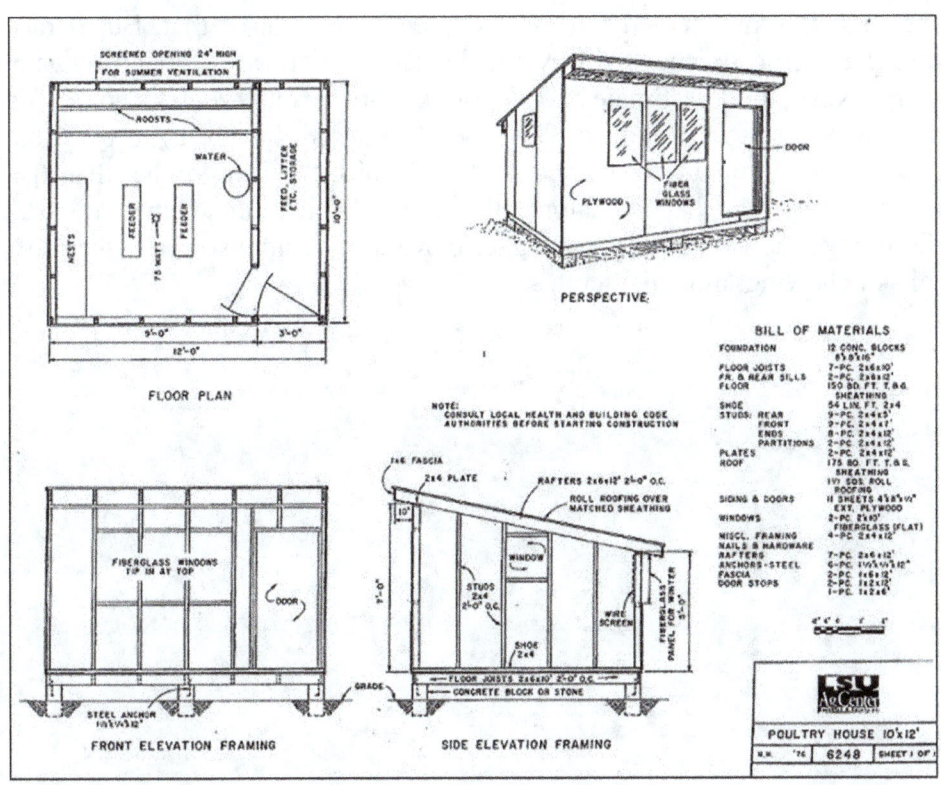

This particular coop is an interesting variation to the above. They are similar in size, with this one measuring 10' x 12', and the previous plan is 8' x 12'. This one has two large roosts rather than a number of small ones, which can make handling birds at night a little easier. The birds can also pack together closer on cold nights. This one also has a small storage room which can be a wonderful place to safety store feed and other supplies, immediately within reach yet away from the birds. This one has fiberglass windows in the front rather than screening, so it would work in colder climates. Additionally, this one makes good use of plywood, which wasn't used in the previous two plans. Plywood is often cheaper per unit area than other dimensional lumber, so this larger coop could actually work out to be cheaper to build. And finally, this one has a constructed floor. That will add to the cost, but will provide excellent insurance against vermin gaining access.

Caution – the space created under this coop could provide wonderful shelter for vermin. Good fencing will eliminate a lot of that. But if rats tunnel underneath, they will eventually chew their way through to gain entry. Never underestimate the determination of rats when it comes to gaining access to a free meal. Two ways to avoid this are a) to build on sand as we've already discussed, and/or b) raise the coop higher than the above plans, so that predators like cats can get under the coop to discourage the rats. A slightly higher elevation would also make the rats' jobs of chewing, more difficult.

## *Plan #4: 25 to 40 Bird Coop*

*Plans:*
http://www.lsuagcenter.com/en/our_offices/departments/Biological_Ag_Engineering/Features/Extension/Building_Plans/poultry/housing/Poultry+House+25+to+40+layers.htm

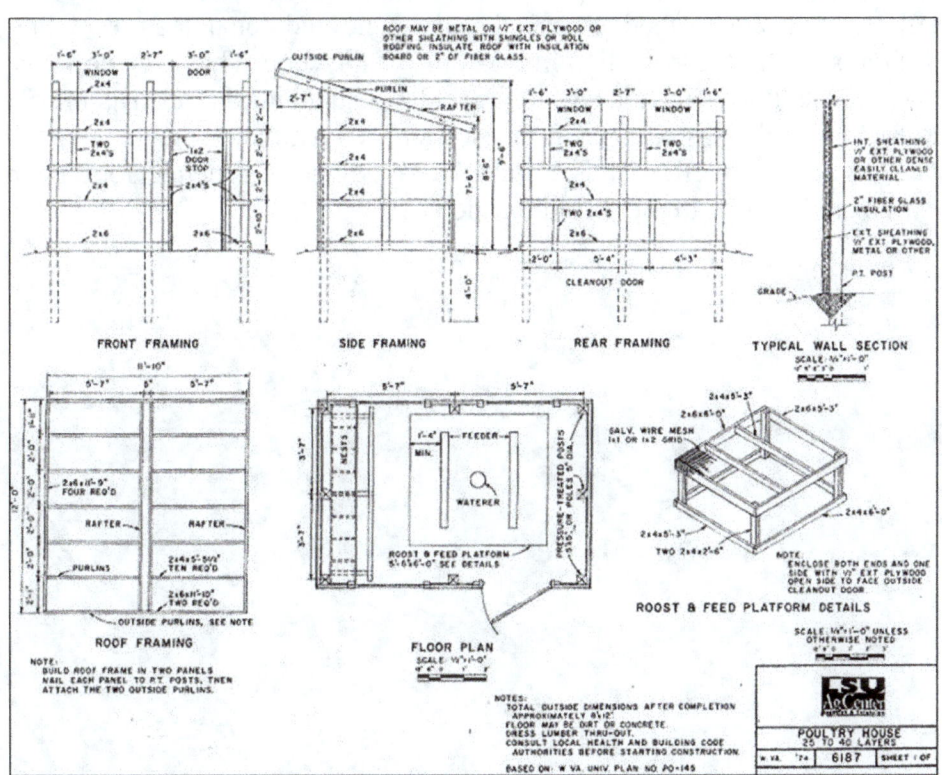

A further variation on interior layout options. This one is back to being 8' x 12' wide, yet is rated to house slightly more birds than the others. This is because of an interesting design feature, where a single platform serves as both a feeder/waterer platform by day, and a roost by night. It's an intriguing idea. One problem with both roost space and feeder/waterer space, is that both spaces are only used part of each day. By combining them, there is room for more nest boxes, and that particular portion of the coop is used all the time. An additional nice feature is that the space under the feeding/roosting platform is accessible from the outside. So cleanup is a lot easier than covering a larger area.

The disadvantage is that if either feed or water is knocked over, it will create a huge mess underneath. So the feeders and waterers will need to be anchored or suspended in such a way that spillage is minimized. Nevertheless, 40 birds in a coop measuring only 8' x 12', is a very efficient and cost effective use of space. One additional feature of this design is the use of insulation in the walls of the coop. This is one of only two designs in this manual, which includes insulation (the other is the 50 to 80 bird shelter listed later in this section). Given the insulation and the nice tight construction, this would be a wonderful shelter for either extremely hot or extremely cold locations.

## *Plan #5: 40 to 50 Bird Small Farm Flock*

*Plans:*
https://www.lsuagcenter.com/en/our_offices/departments/Biological_Ag_Engineering/Features/Extension/Building_Plans/poultry/housing/Poultry+House+4050+Birds.htm

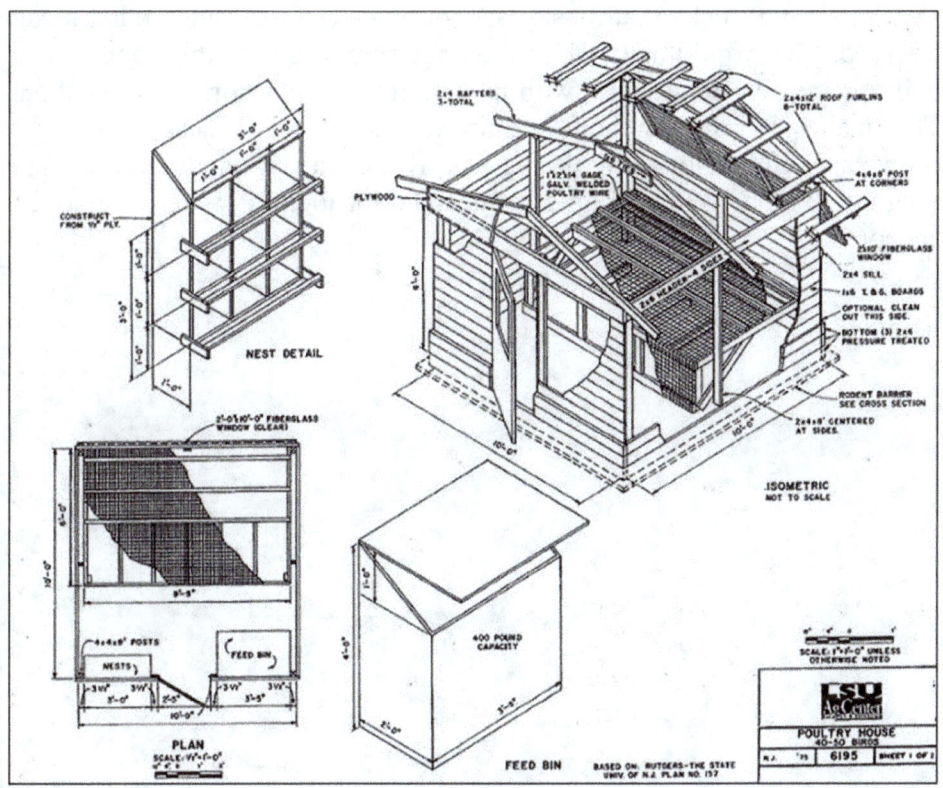

This coop is getting large enough to be out of reach for most suburban homeowners, yet it's a very good starter flock size for homesteads and small family farms who want to sell eggs. This coop is chock-full of some really nice features. First of all, the nest boxes are stacked alongside one wall, rather than taking up valuable floor space arrayed in a row. Birds can easily navigate up to stacked nest boxes, and they should be used more often when space is at a premium. Secondly, this has a nice full-width fiberglass window for plentiful light in winter, yet it rotates open in summer. This also has a roost platform, with a cleanout space below. And finally, this coop has a built-in feed bin to dispense feed over time for a large number of birds. This reduces the amount of daily work to be done by the flock owner.

Several notes of caution with this coop. First, it does not specify whether a cleanout door is along that back wall, such that the roost space can be easily accessed. Make sure that cleanout door is included or cleanup will be a hassle. Or, make that roost platform hinged to the wall, so that it can be lifted. That would make cleanup very easy. Secondly, this coop is also built directly on the ground, with no apparent soil anchors or foundation. This makes it prone to both vermin access, as we've seen with other coops, as well as blow-over during stormy weather. If winds ever surpass roughly 35mph in the coop's immediate location, it will need to be anchored.

## *Plan #6: 50 to 80 Bird Small Farm Coop*

*Plans:*
http://www.lsuagcenter.com/en/our_offices/departments/Biological_Ag_Engineering/Features/Extension/Building_Plans/poultry/housing/Poultry+House+50+to+80+Layers.htm

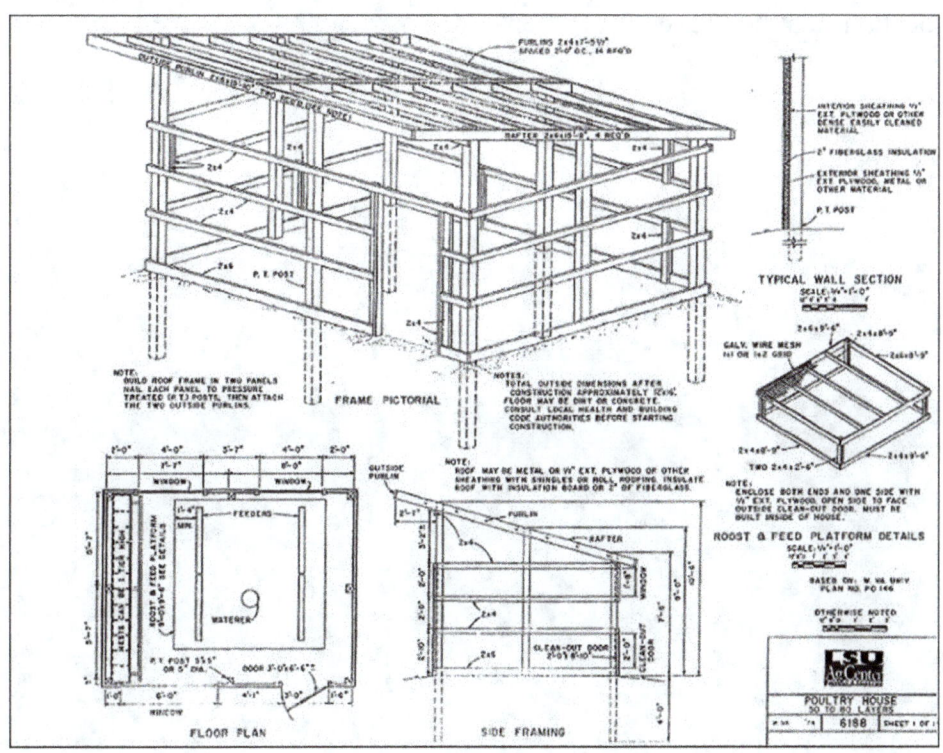

As with the previous coop, this coop is well beyond the size needed by suburban homeowners, yet well within the flock size for homesteads and small farms. This coop is slightly larger in most ways, with a larger overall footprint (12' x 16') than we've seen prior. It has the same combination roosting/feeding platform, against the back wall as before. This also has the nest boxes mounted against the front wall, with the option to stack them vertically for as many as 24 total. The main difference is that this building as such, offers a relatively large free-span area inside. Limber dimensions will have to be followed closely so that the roof does not sag under load. Also notice that the corner and perimeter posts are sunk well into the ground. This anchoring provides crucial strength given the lack of interior support.

As we've seen with other designs, this is built directly on the ground, so is at risk of vermin tunneling under the walls. Flock owners can either opt to install a floor or use a sand base. And as mentioned earlier, this design is one of only two which uses insulation. Given that, it would serve nicely in locations that need extra protection.

## *Plan #7: 100 Birds Farm Flock*

*Plans*:
http://www.lsuagcenter.com/en/our_offices/departments/Biological_Ag_Engineering/Features/Extension/Building_Plans/poultry/housing/100+Unit+Laying+House.htm

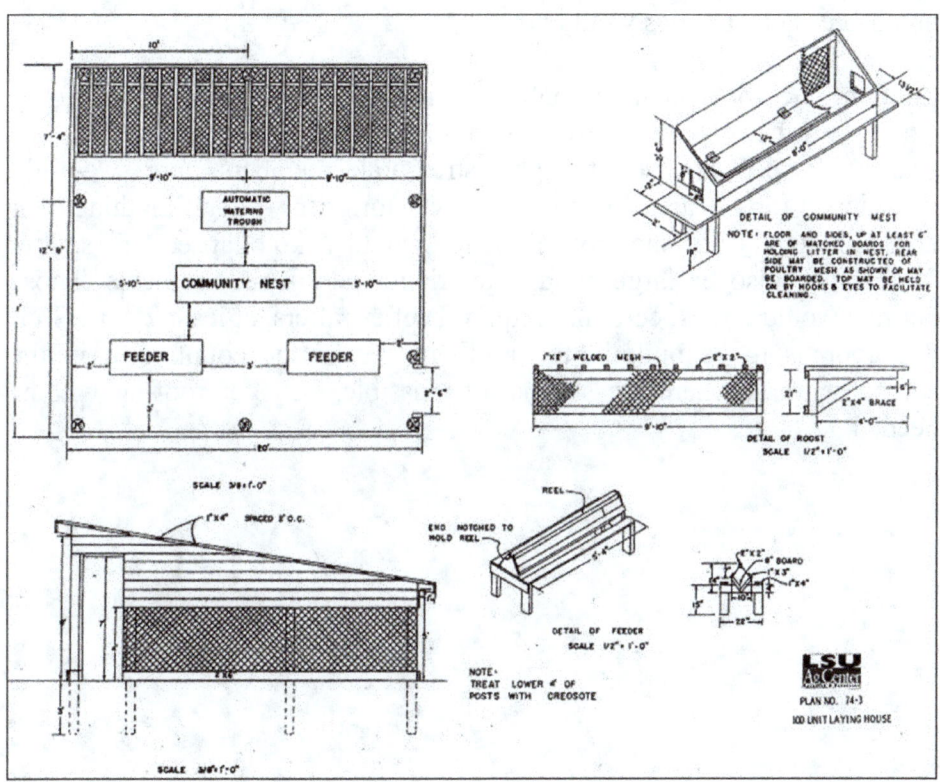

This is the largest coop we'll be looking at from the LSU AgCenter collection. A flock size of 100 birds is a solid flock size for small farm egg production. A large homestead or small farm can build this coop and grow into it, knowing that both present and future flock needs will be provided. This design makes use of a few nice features. First of all, plans call for an automatic waterer. This would be a nice feature but only if well protected from freeze damage in winter. Check this manual's section on feeders and waterers for some ideas on that topic. Secondly, we see another community nest here, but this time it's situated right in the middle of the pen, between the feeders and waterers. That layout is convenient for the birds, and may be convenient for the flock owner, if the flock owner sets up chore procedures to take advantage of that placement. The roosts are all against the back wall with a nice contained manure spot underneath. Finally, this design features plans for a homemade bulk feeder. While this coop is intended for chickens, it would also work for a small turkey operation. Heritage breeds of turkeys would need the roost space, but broad-breasted varieties would not.

Caution: the roof pitch is shallow, with relatively small rafters (only 1"x4"). Furthermore, the load is carried by the outside posts alone. The shallow slope, and lightweight structural members, are woefully insufficient for any areas with snow or ice storms in winter. Anything over approximately 1" of snow or ½" of ice would risk collapse. This shelter design may also be large enough to trigger permit requirements. Those permits would almost certainly require beefier rafters, at least 2"x6". So if this coop is to be built in the northern tier of the country, or in the southeast areas where ice storms are possible, stronger roofing will be needed.

## Plan #8: Field Shelter for 100 Bird Farm Flock

*Plans:*
http://www.lsuagcenter.com/en/our_offices/departments/Biological_Ag_Engineering/Features/Extension/Building_Plans/poultry/housing/Summer+Range+Poultry+Shelter.htm

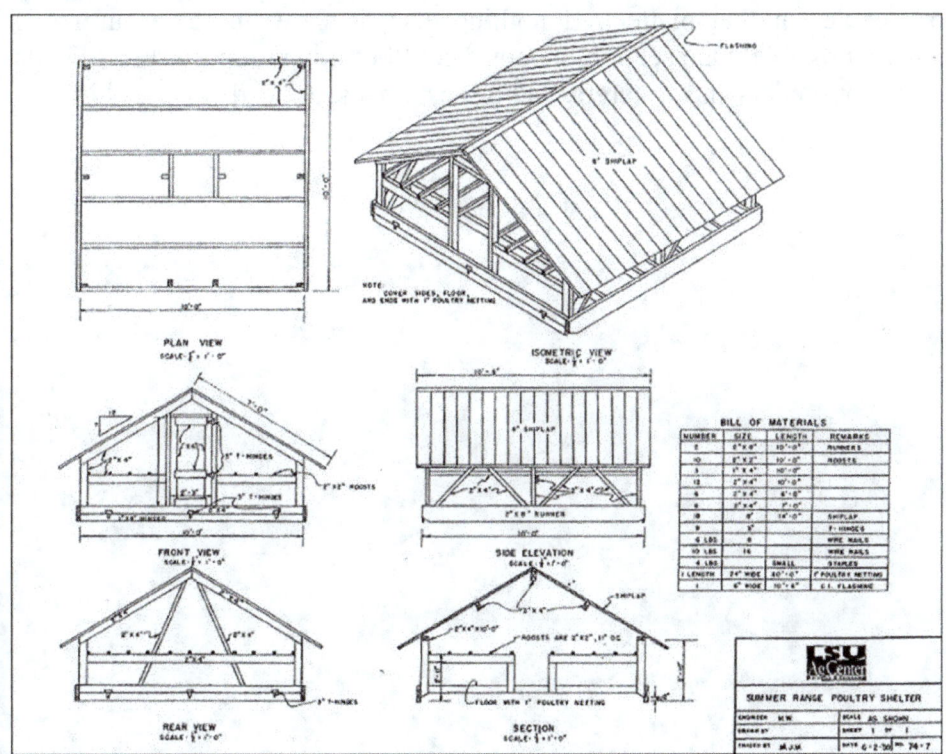

This is one of the few stud-construction mobile coop plans I was able to find for this manual, and the only free set. While considerably larger than suburban, small homesteads or small farms would need, it gives some nice ideas for how to build such a mobile shelter. The shelter design could simply be down-sized for smaller flocks, or the existing design used to either provide more ample cover, and/or provide for larger future flocks.

This shelter is essentially one big roosting area. Nest boxes are not specified in the shelter plans, and would need to be added for a layer flock. Additionally, feeders and waterers are also not provided with this structure. So they too would need to be added, or available elsewhere. As such, this structure doesn't really qualify as being a chicken tractor, since the birds are free to roam away from it, and it doesn't provide for all their needs. Furthermore, most modern field shelters are made from substantially lighter-weight materials than the wood construction pieces specified here. I include this model here to show a rugged wooden design, set up to be moved via tractor (or horses or ATV or SUV or even a group of people for that matter). If nothing else, it shows us an example of what's possible. One section of roosts could easily be removed to provide room for feeders and waterers, and/or nest boxes, as needed.

# SECTION III: ALTERNATIVE COOP DESIGNS

*A group of four hens on top of a fence railing. They are alert and lively this foggy winter morning despite temperatures near freezing. Healthy adult birds can withstand very cold weather if they have protection from wind and precipitation. Photo by author.*

In the previous section, we reviewed a variety of coop plans which all used conventional stud construction. Most of them, for most people, would easily qualify as a nice range of chicken coops. Yet this manual has two goals – describe both conventional, and unconventional, coops. Let's look now at unconventional coops.

As we did in the last section, we'll start this section with a few assumptions that the flock consists of:

- Chickens of any breed, rather than waterfowl, turkeys, pheasant or other game birds.

- Most of these birds are adults, thus don't need supplemental heat.

- Layer coops, but several of these are already in use for broiler pens.

- These birds will have access to an outside yard of some kind.

By using the same assumptions in this section as before, we can start to look at alternative designs that can be compared side by side with the others in terms of how many chickens can be so housed. I will say here that many of these designs offer some major advantage over stud construction, but typically will also have some major disadvantage. There's no such thing as a perfect design, and there are almost always tradeoffs. I'll try to call out those advantages and disadvantages with each design, but each flock owner will have to decide for themselves whether those advantages and/or disadvantages add up for his/her own particular circumstances. So let's get started!

## *Car Canopy*

*One of the author's very first chicken coops. It provided basic shelter, but had a variety of issues.*

Many home improvement stores, auto supply stores, and online retailers offer variations on this theme. The theme is a metal pole framework with a fitted tarp stretched over the top and sometimes the sides as well. While most car canopies are considerably taller than the photo below, that situation is easily amended by simply removing one set of leg sections.

The above is a photo of one of our first chicken coops, from approximately 2002. The footprint measured approximately 10'x20', and easily housed up to about 50 birds. We bought this shelter as a 3-season shelter only, but somewhere along the line it turned into a year round structure. We used it for almost seven years, through high winds, heavy rain, occasional snowfall amounts up to 12" at a time, and all the other weather variations which affect us in our Pacific Northwest location.

Let's take a look at the pro's and con's of this design, with the con's first. For anyone who has been reading this manual closely up to this point, several glaring problems will immediately be apparent:

- No protection from rodents

- Lightweight protection from predators

- No protection from strong winds

- Minimal protection from heavy snow loads

- No protection from mud

Indeed, we have had problems with all the above over the passage of time. The first and last issues, rodents and mud, have proved to be the most difficult to deal with. We've been dealing with them ever since. As we've logged time and experience with poultry, we have found ways to minimize those problems and/or improve this design's weaknesses. Specifically, we've made the following changes:

- Heavier tubular frames to better withstand wind and snow loads.

- Anchoring the structure to the ground so winds won't disturb it.

- Running lines over the top of the tarp to keep it from flapping against the frame.

- Using natural vegetative shelter to reduce side winds.

- Reducing the overall height, to reduce the exposure on each side.

- Cats living in the chicken area to hunt rodents.

- Hotwire fencing around the perimeter to reduce predators.

- Livestock guardian dogs around the perimeter to further reduce predators.

- Use of plentiful bedding to deal with mud issues.

The above improvements have reduced, sometimes eliminated, the disadvantages of this design. Enough so that we've continued to use this design over nearly a decade of flock ownership.

- Now that we've looked at the con's, let's look at the pro's:

- Low initial purchase cost

- Extremely easy assembly

- Extremely easy relocation

- Very well suited for mild conditions and/or sheltered locations

When we first got started with chickens, we weren't sure how things were going to go. We didn't want to invest a lot of money into a coop only to learn that we didn't like birds. And we certainly didn't think far enough ahead to see all the above problems looming on the horizon. So we went with this design as our first coop. Despite the disadvantages listed above, we kept using this particular shelter, and a few more like it, for many subsequent years. While we were frustrated by the problems, the advantages continued to justify continued use.

Overall, it's been a serviceable design which obviously has worked "well enough". Yet as those shelters are wearing out, we are planning to replace them with other structures which will avoid the disadvantages, and capitalize on the advantages. I'll speak more about that next.

## *Greenhouse/Hoop House*

*The winter housing for Kent Solberg's layer flock at Seven Pines Farm, Verndale, MN. Photo taken by Adrienne Cahoon, Whiteley Creek Homestead Bed & Breakfast, Brainerd, MN. An excellent example of how light, airy and roomy the interior of even a small greenhouse can be for birds. Visit Adrienne's blog write-up of her tour of Seven Pines Farm, at http://www.whiteleycreek.com/queen_of_the_meadow_bloom/2009/05/ .*

This design is also quite common, particularly amongst small farm producers who already have hoop house and/or greenhouse experience. Hoop houses and greenhouses are now so easily acquired, or built from scratch, that many folks can readily find materials and/or specific designs of various sizes.

In the photo above, the greenhouse is used for wintertime housing. In summer, the birds are out on pasture. At the time this photo was taken, the birds had not yet been moved off pasture for the year, making it easy to see the interior layout. Additionally, a nice bed of either sand or shavings is also visible, as is the nice combination of rigid siding along the base of each wall. This is a wonderful environment to raise healthy, happy, clean birds without mud, rodents or other issues. This is the design we're working towards having.

The list of advantages continues. This structure is very stable in wind, and will readily shed rain, snow and even ice without the interior getting wet. If the flock owner desired, this structure could be either heated with off-the-shelf greenhouse heaters, or cooled with either active or passive ventilation systems in the walls. For climates with hot summers and cold winters, the dedicated flock owner could do both. Additionally, a small flock could share this very nice growing space with plants for either seasonal or permanent growing needs. In that instance, a partition would be required so that the birds didn't get into the plants and do damage. Yet this design is flexible enough to do both. Alternately, a greenhouse could be used for summer plant production while the birds are on pasture, then shelter the birds over winter when the plants are dormant. Many, many options here. The structure itself will survive for decades if it's kept clean and rust-free. The greenhouse film will require more frequent replacement, but even there each covering should last at least several years. The furniture is rugged as well, and should provide years of trouble free service. This is a coop ready for use.

One final advantage is the concept of different housing for winter vs. summer. The above greenhouse is used for birds during the winter, but they are out on pasture during the summer. Their summer housing is portable, lightweight, quite small and provides only a minimum interior for nesting and roosting. The birds are out on pasture most of each day so the shelter doesn't need to provide much else. Yet in winter, the birds need a much larger interior space to allow for normal social interaction and exercise during inclement weather. That mobile field shelter would never be able to provide that; this greenhouse provides it very well. Even better, this greenhouse could potentially be used for other purposes while the birds are on pasture, thus getting two valuable buildings for the price of one. Greenhouses and hoop houses are so very versatile that way. If a flock owner puts up a greenhouse or hoop house for winter poultry needs, he or she will almost certainly also find a way to use that same shelter during summer.

The only disadvantages are cost, and complexity of construction. This is a substantial building, and its cost and construction will reflect that. However, that is not necessarily written in stone. Not all hoop houses and greenhouses are as rugged or permanent as this one. We have hoop houses here made of outdoor grade PVC tubing, which are extremely lightweight and portable. Some greenhouses are made of wood framing with gable roofing to approximate the curve of the metal hoops. And some are specifically designed to be dragged with a tractor across the terrain. This is to provide coverage of alternating or sequential areas, with only one building. Ironically, the designs which are intended to be dragged are often heavier than their permanent counterparts. This is because of the additional reinforcements needed to resist distortion during movement. Any of these designs will work for flocks, depending on the flock owner's priorities. And greenhouses and hoop houses are now common enough that anyone interested in such a design can find plentiful local suppliers and online retailers.

## *Truck Bed With Canopy*

*This is a photo of the author's compact pickup with canopy shell, temporarily set up to serve as poultry housing. It would not be ideal if this vehicle was used for regular errands. However, it can be a very tidy short-term solution, and a nice option for a vehicle which is no longer roadworthy but can still get around.*

This shelter method is generally used when a flock owner already has a non-functional truck and/or canopy ready for use. The canopy by itself can certainly be used as a non-walk-in shelter. Most canopies are easily tall enough to very comfortably house a small flock, say up to 6 birds with one roost, 12 birds if 2 roosts can be fitted within. If a canopy were mounted on a truck bed, interior height is greater and the birds would have plenty of space. A small feeder and waterer could also fit inside, as could a nest box or two. And finally, if the truck and/or the canopy are already available and not being used, well, a free shelter is a free shelter, and it beats the birds roosting up in the trees.

This housing would be about adequate for up to 6 birds. It offers a single roost towards the front of the truck bed. That roost is just visible above the top of and behind the kitty litter box. Note the hay bedding immediately underneath the roost. The litter box, meanwhile, is serving here as a nest box. A single feeder and waterer complete the setup. If we really needed the space, this would definitely serve. Since we already had the truck and canopy, the cost of the remaining items is almost negligible. Even normal household items like the kitty litter box and the roost pole (a repurposed leaf rake handle) could be used rather than buying in chicken-specific furniture.

Besides cost, this setup offers some surprising advantages. First of all, it's almost predator proof, at least in terms of the space within the truck bed itself. Birds would need to be let out into a yard or pasture during the day for light, sun and exercise. That enclosure would be more or less predator proof depending on how well it was fenced and what other shelters may be available. This enclosure is also relatively rodent-proof. We've very occasionally seen rodents get up into a truck bed when we've had something stored in the bed for long periods of time, such as bags of feed. But they sure have to work at it. And finally, this coop is as mobile (or not) as the vehicle it's housed in. If the truck is in service, it can be driven to wherever the chickens are to be grazed for the day. That beats dragging around a heavier trailer or other structure.

The major disadvantages are the height and access. This would typically qualify as a non-walk-in coop, thanks to the lack of interior standup height. Larger canopies would allow a person to crawl in to gather eggs, refill the feeders and waterers, and otherwise check on things. But that's hardly desirable. Even if the person didn't try to crawl in, full access is tough. The truck bed walls and the canopy walls are typically built strong enough that access is difficult or impossible without extensive, and expensive, modifications. Some canopies have side windows which can be lifted and raised from the outside, specifically to allow people to reach in from the sides. Other canopies like the one above have slider windows which cannot be opened from the outside. Yet if the interior space is organized carefully, that isn't necessary. For instance, if the feeder, water and nest box are located next to the tailgate, they are easily accessible for daily tasks. It makes sense to put the roosts towards the front of the vehicle, which is easy for the birds to reach but difficult to clean. A long-handled rake could be used to rake out the bedding under the roosts. Not the most convenient, but definitely do-able.

Sometimes, either due to emergencies or financial shortages or whatever, a truck bed and canopy could serve very nicely if nothing else is available. As my father is fond of saying, "any port in a storm." If birds need an emergency shelter, a truck bed and canopy can provide that shelter. They'd be very comfortable, even if tending them would not. Short term, that wouldn't be a problem. Long term, some other solution would probably be needed.

## Shipping Crates and Containers

*This compact little chicken coop was created out of a pair of shipping crates by the Old Tyme Coops company. The coop's lid, visible in the lower left of the photo, is about to be installed. More information about this coop design is available either from the company's Facebook page, https://www.facebook.com/pages/Olde-Tyme-Coops/487040134690970, or from a photo montage of the coop's construction visible at http://www.backyardchickens.com/t/746561/shipping-crate-coop.*

Shipping crates and containers are possibly one of the least-well-utilized, commonly available materials, in modern society. We tend to disregard them as "trash" or as waste products, given that they are merely protection for some more valuable commodity within. But they are often extremely rugged, inexpensive (or free), and they're available everywhere in some form. The form can vary anywhere from small reinforced wooden frames with or without plywood or chipboard sheathing, all the way up to the monster metal 10'x40' shipping containers used for trans-oceanic product movement. The latter in particular have become very popular on the used market, and a number of companies have sprung up specifically to buy these, perhaps recondition them a little for some particular purpose, then resell to consumers. Other forms of crating are only locally or seasonally available, and a person needs to pounce on them when they're made available. Any of these containers can be remade into deluxe poultry housing, with a small to moderate amount of elbow grease. And often for a fraction of the cost of a similarly-sized, conventional stud construction building. So let's take a look at the possibilities.

The advantage of this approach is mainly cost. Many used shipping crates can be picked up for dramatically less cost than the same materials would cost new. Sometimes they are available for free. Many times, an enterprising and creative flock owner can proactively contact nearby appliance dealers, manufacturers and/or wholesalers who regularly get crates as part of their product shipments. These companies often have to dispose of these used crates at some cost, so they are eager to find other uses for them. If a steady supply can be found, that resource can converted into coops for sale. That's ultimately what happened with Olde Tyme Coops.

The major downside of this approach is that the flock owner needs to carefully consider how to make the most of the existing crate, yet still achieve the designs needed for the flock. Sometimes the available crate or container needs such extensive remodeling that the cost advantage starts to disappear. Alternately, the design which makes the most of the available container, imposes limits which will become a long-term annoyance for the flock raiser. or instance, metal shipping containers offer a large, robust, easily cleaned surface which is rodent-proof, predator-proof and weather-proof. Yet ventilation can be a problem, resulting in condensation or ammonia build-up during the winter and heat build-up in summer. Increasing ventilation would require cutting into the metal siding of the container, which then requires either a very robust metal cutting blade, or a cutting torch. And once those cuts are made, they can't easily be unmade. So careful consideration of how to optimize the design given the container, and the flock, will be a flock raiser's best recipe for coming up with a winning, workable design.

## Horse or Livestock Trailer

*The above photo is a two-horse slant-load trailer, owned by a neighbor, converted to be a short-term chicken coop. Photo taken by author. The afternoon sun streaming in the windows helps illustrate how open and airy this design can be. This trailer will have good light and ventilation even when the two back doors are closed and latched.*

Sometimes, you use what you have. Many folks who have other forms of livestock, will have a horse or livestock trailer which can be pressed into service as a chicken coop. While this idea may seem rather nutty, it's actually quite ingenious. Used trailers can serve as either a temporary or a permanent field structure for pasture-raised birds. Almost any size horse or livestock trailer is nearly ideal for this purpose because it's already designed to be mobile, easily cleaned, lightweight yet rugged, and provide a nearly ideal balance of shelter from the elements with sufficient light and ventilation for the occupants.

If a flock owner already has a livestock trailer, it can be converted relatively easily. The above horse trailer, owned by a neighbor, is typically used only a few times per year for moving her horse. Conversion from being a horse trailer to a chicken coop took me a total of 15 minutes. I simply spread around some wood shavings, put in a feeder and waterer, and used a clean kitty litter box as a nest box. If I really wanted to outfit this for regular use, I could have fastened a roost bar ladder along one side. This is a slant-load trailer so the front wall was at an angle and thus not very conducive for mounting the roost. If it had been a straight-load trailer, I could have put the roost bar(s) towards the front, so they wouldn't be in the way of daily cleaning and feeding duties. So if a flock owner already has a trailer which sees occasional or even rare use for moving large stock, it can serve double-duty as a temporary coop. Better yet, if a trailer is really worn out, with a bad floor, perhaps a not-worth-repairing frame, anything which makes it non-road-worthy, it can still serve admirably out in the fields for the birds. Many a trailer has been retired to this lightweight purpose. Even the undercarriage can be pressed into service as the base of a mobile coop, if the rest of the trailer has rusted away or is somehow damaged beyond repair. Similarly, if the topside is in good shape but the undercarriage is hopelessly mangled, it can be parked or even mounted on skids out in the field, and still work nicely as a coop.

However, if a flock owner does not yet have such a trailer, some disadvantages should be considered. First, the size will become a consideration if the flock owner wants a flock size beyond 12 or so birds. After that point, the trailer will quickly become too small. Secondly, trailers are designed for horse and/or cattle dimensions. The heights and widths are ideal for those species. They are not designed for birds, or humans for that matter. As a result, human flock owners may find the space a tad too constricting for easy chores duty. Taller flock owners may even bump their heads against the roof if the trailer is particularly small, and/or has interior bracing between the walls and roof. And finally, if the trailer's tires are not maintained and kept aired, they will eventually go flat at which time the trailer is no longer mobile. Acquiring a used trailer may be quite easy in rural areas, and quite a challenge in suburban or urban areas. So location will be an issue. And finally, sometimes trailers need to be licensed even if they are not intended for road use. That varies by location as well. As with any other form of shelter, flock owners will have to balance these pro's and con's.

## Cattle Panels and Tarps

*Photo of three broiler pens constructed from cattle panels, framed out with wood, and covered with inexpensive blue tarps. Photo courtesy of Kinikin Heights Natural Foods of Gunnison, CO, http://kinikinfoods.com/.*

Cattle panels, also known as hog panels and livestock panels, are all forms of stout fencing. They are large gauge rigid wire, welded into a grid pattern, typically in sheets which measure either 16' or 20' long, by 3', 4' or 5' wide. They are generally used to form small paddocks for either cattle or hogs, as the name implies. However, quite a few thrifty farms, ranches and flock owners have used them to create small shelter frames.

The gorgeous photo above is a nice display of cattle panel construction. The pens are framed with wood along the bottom, and have cattle panels arching over the top to form a weather-resistant roof. Cattle panels also form the front and back to provide either full or partial protection. The wood framing at the bottom keeps the footprint stable, so the arch is maintained even as the pens are moved. These shelters also demonstrate how flexible standard blue tarps can be, providing either full protection against sun, wind and rain, or opened up to allow more exposure on nice days. These shelters may not be up to harsh winter conditions in many areas, but they would be perfectly suitable as field pens during the growing season.

The main advantages of this design include the flexibility of design in terms of how much area is open versus sheltered. That type of feature turns this into a shelter that is very easy to use throughout the growing season, regardless of all but the harshest weather. Furthermore, it's lightweight yet rugged, with reinforcements at key locations without being overbuilt in general. It's nearly predator proof, with the only risk being from predators burrowing under the edges. And it was less expensive to build than a conventional stud construction building would have been, particularly given the square footage provided.

I should point out one additional advantage. In heavy rainfall events, this design is amongst the best available in terms of preventing sags in the roof. Conventional greenhouses must stretch the film very tightly across the frame, and/or inflate the frame, to keep the sheer weight of rain from ponding on the roof and thereby stretching the film. Tarps stretched over metal frames, as in the car canopy model discussed previously, cannot be stretched tight enough to prevent that ponding effect. Those sags become so large over time that they allow a lot of weight to build up – sometimes enough to collapse the frame. Yet the grid pattern of the cattle panels creates a near-uniform surface, which results in much better shedding of rain. The only roof which performs better in this regard is conventional rigid roofing as part of stud construction. This solution is dramatically less expensive.

This particular design would not stand up to a heavy snow or ice load. If it is used only seasonally, it may never need to. However, this basic design could be turned into an all-season shelter with the addition of a central ridge beam and beam supports at either end. The footprint could also be narrowed a bit so that the sides sloped more, and met the ground at more of a perpendicular angle. That would ensure that snow didn't pile up along the lower section of the roof and weight it down. But these are very simple modifications to a generally very workable design.

The only other two disadvantages are also minor. First, rodents or predators could eventually get into the enclosed area by tunneling underneath the side walls. This may or may not be an issue, particularly if these shelters are frequently moved on pasture. Secondly, a shelter like this must be moved frequently to ensure the birds are on fresh pasture. This is particular true with a growing population of broilers inside. At first, the pens would only need to be moved once every few days for younger birds. Towards the end of the grow out period, a pen of this size with this broiler population may need to be moved twice a day. This would ensure fresh grass and clean birds. That is standard for field pens of any kind, and this pen is no exception.

In any case, this is a beautiful example of how cattle panels can be used for providing at least seasonal shelter. Any of the disadvantages listed above could easily be dealt with by tweaking the design only slightly.

## Metal Erector Frames With Tarps

*A gambrel-roofed animal shelter made from metal EMT poles, joined with metal slip-fit connectors, covered by a heavy-grade tarp. The connectors and tarps are available from www.shelters-to-go.com. Photo used by permission from www.shelters-to-go.com. The entire construction project is documented on their website.*

This building method is very similar to the car canopy method described previously. So similar that I debated whether to include it separately. I decided to include it because some options are available with this approach, which are not available from pre-made car canopy kits.

The above photo was from www.shelters-to-go.com, and it showcases one of their more unique products. Shelters-to-go and other similar businesses have created a niche for themselves alongside car canopy and other shelter manufacturers. Rather than selling complete kits of specific size, Shelters-to-go and others like them sell the joints, or "connectors" used to match up and align metal EMT poles into three dimensional shelters. The EMT poles themselves are commonly available at hardware stores and big-box home improvement centers, sold as conduit in the electrical section. These poles are generally available in 10' lengths in a variety of diameters ranging from ½" to 2". They can be cut down to any smaller lengths. The genius of this combination, is that structures of any size and roof slopes of different angles, can be assembled on-site per any specific needs. While flat, shed-type or gable-type roofs are the most common, gambrel-type roofs like the above are also possible. The above shelter is intended for goats, but the same shelter would provide admirable shelter for a large flock of birds. Scale down any of the dimensions and the shelter could easily house smaller flocks.

The advantages of the above shelter include cost, ease of construction, good ventilation, and better than average mobility. The EMT poles and the connectors themselves will be the most expensive portion of this construction. The tarp, or similarly-sized greenhouse fabric, will also be a single large purchase. Thankfully, different grades of tarps and different greenhouse film thicknesses provide some leeway in that portion of the budget. Yet these structures can span such large areas so efficiently that they're still cheaper than stud construction. Ease of construction bears some additional comment. Stud construction takes time in part because there are so many pieces – the stud frame by itself can be composed of dozens if not hundreds of individual pieces of dimensional lumber. Add to that the need to cut angles for things like rafters, and nail each piece of dimensional lumber multiple times. Building an 8'x8' stud construction shed would take a whole weekend, and that's if a builder is experienced and organized. Yet the above shelter will go up in a fraction of that time, because there are fewer pieces to handle, and no complex angles to cut. The connectors work by compression screws, so nails aren't needed. Tarps or greenhouse film would need to be well anchored along the sides, but even that is faster than standard plywood sheeting roof construction.

Ventilation can very easily be controlled by the degree to which the ends are left open, or closed off. This doesn't have to be an all-or nothing design. The photo above shows the back end closed off by a 4' high wall, and a single large custom-fitted tarp. The wall would have taken awhile to construct but the tarp would go up in minutes, and can be taken down just as quickly. The front portion of this shelter, not visible in the photo, is completely open. This too could be closed off, or opened up, as needed.

This particular shelter is not mobile because of the plywood base walls, but that was the choice of that particular builder. The metal frame itself is lightweight enough that it could either be pulled along the ground by a tractor, ATV or SUV, or even a team of people. Alternately, a wood platform floor could be built and the metal frame mounted right on top. Then the whole building could be pulled along like a sled.

One interesting aspect of this approach is that the strength of the structure can be customized according to local needs. For instance, we started off using 1" connectors and rods for our goat shelter, about 10 years ago. The building has done pretty well but I need to sweep the roof off when it accumulates more than 6" of snow. So a compost roof we built a few years later used 1.5" connectors and rods. That building had the same footprint and the same overall construction method, but it's more resistant to snow. If wind and snow aren't an issue, a smaller diameter would save money.

The downsides to this approach: zero resistance to rodents, and only slight protection from predators. Tarps and greenhouse film are wonderful visual barriers but they pose no protection at all against claws determined to get inside. It wouldn't be instant, but anything the size of a cat could work through a tarp given enough time. Even a rodent could chew through it. The wooden base walls in the photo above would prevent most of that, but then mobility would be lost. A wooden platform on runners would offer both predator and rodent protection, while preserving mobility.

The only other disadvantage has been mentioned before. Heavy rainfall, snow or ice loads will slowly but surely create enough weight on the tarp or greenhouse film, that pockets will form and allow pools to gather. Those pools, over time, will enlarge and become a weight issue. For flat or shallow-slope roofs in high-rainfall or high snowfall areas, this deterioration may occur within a single year or two. For steeper roofs, that process will be slowed considerably but construction costs will be higher. For gambrel roofs like above, only a small fraction of the roof would run that risk, but it's also the highest portion of the roof and thus the hardest to reach. For that reason, some additional protection should be used during construction, such as the use of cattle panels. See the cattle panel section for a discussion on how that material helps combat this problem.

This approach offers a lot of flexibility and cost effectiveness, and they are gaining popularity over cookie-cutter car canopies. As long as flock owners really consider their particular needs for wind and snow resistance, they can size the building materials accordingly and get a very cost-effective shelter.

## Pallets

*A pallet coop designed and built by Butch Bridges, of Lone Grove, OK. Photos taken by Mr. Bridges. The entire project is detailed on his website at http://www.oklahomahistory.net/chickencoop.html.*

Pallets, like shipping crates, are considered a throwaway item by modern society. Yet pallets offer extremely sturdy panels, of typically high quality lumber. If they are cheaply or freely available, they can be used very successfully as coop building materials. Just add some creativity and elbow grease.

The above series of photos show just what is possible when using pallets to build a larger structure. These are only three of the photos of that project. Many more photos showing the entire project are available on his website at http://www.oklahomahistory.net/chickencoop.html. While this particular coop probably went where no other pallets had gone before, and the builder obviously took this concept well beyond mere necessity, other flock owners can use pallets or pallet material for a wide variety of designs.

Pallets offer two basic options. First, the pallets themselves can be left intact, and used as panels to form some overall structure. As the above photos demonstrate, multiple pallets of the same overall size and dimensions can be connected and used to form a larger uniform space. The second option is to dismantle the pallets and use their dimensional lumber components. Different pallet manufacturers offer slightly different dimensions and shapes for those components, so the prospective builder needs to examine the pallet supply carefully in advance.

One of the best potential options if pallets are going to be used, is to proactively gather as many identical pallets as possible, and then use them as panels as illustrated above. I say this after dismantling dozens, if not hundreds of pallets over time, and trying to work with those individual pieces. There are definitely designs which would make good use of those components. If a flock owner has a specific design in mind, by all means pursue it. Just be aware that what money is saved by gathering the pallets, will be lost in the time it takes to dismantle them and reassemble them in some other arrangement. For that reason alone, I recommend using them as panels.

When used as panels, the enterprising flock owner and builder can make the coop up to approximately 16' long, up to 12' wide, and up to 4' high, before pallet connections and structural reinforcement really becomes an issue. The coop in the photos is near that limit, with an overall length of about 12', a width of about 4' and a height of about 4'. Those dimensions are determined by the width of the pallets, and how they are connected. This is a situation where the building materials will dictate what shape and connection form is easiest to work with. Just be aware that these dimensions happen to work quite well for birds, while they don't work as well for humans who need to work within that space. No one wants to crawl on their knees into a half-height space to refill feeders or gather eggs. So whatever dimensions or construction methods a builder chooses, be sure to include some provisions to keep the coop easy for humans to work with. One simple solution, as in the photo above, is to raise the whole thing up on stilts or legs. The interior space is still plentiful for the birds, while human caretakers will have a much easier time with daily tasks.

It's hard to list additional advantages and disadvantages, because those depend entirely upon the specific design to be used. To see some of the wide range of options, Google the keyword phrase "pallet chicken coop" and look through all the images that come up. Those will give some idea of the variations possible with this approach. If a blanket statement can be used to cover all those variations, the main advantage would be plentiful building materials for very little or no cost. The main disadvantage is that the flock owner has to design the coop very carefully to make the best use of those materials. I generally encourage people to go for it if they have any interest in doing so. Just remember to follow the age-old adage: "measure twice, cut once".

## *Rebar Frame*

*This lightweight, portable yet rugged field pen is used at Sugar Mountain Farm of West Topsham, VT for their laying flock. Photo taken by Walter Jeffries of Sugar Mountain Farm, and is used by permission.*

Rebar is amazing stuff – strong, lightweight, readily available, easily worked, and once it takes a shape it'll try to keep that shape. Any such cooperative material would of course eventually be drafted into use as material for a chicken coop. So let's look at how that's done. The above chicken coop is one of several used at Sugar Mountain Farm for their pastured layer flock. A more detailed description of this structure, and this approach, is available at http://sugarmtnfarm.com/2009/03/08/chicken-hoop-house/.

I have only seen a few coops using rebar, but I generally liked what I saw. To be specific, most of the coops would more accurately be called chicken tractors, because of their size, mobility and purpose. As the above photo shows, such coops are typically quite small, and intended to be moved frequently. They are often covered by chicken wire, tarps, hardware cloth or similar wire mesh, or some combination like above.

The basic premise is fairly straightforward. The rebar itself is used to create a lightweight yet strong frame. The size and shape of that frame is determined by the length and diameter of rebar used, and how it is joined. Rebar lengths are most commonly available in 2', 3' and 4' lengths at hardware stores and home improvement centers. Specialty retailers offer them at longer lengths. Since most folks wouldn't conveniently have access to such specialty suppliers, the shorter lengths are more commonly used. As such, that limits the size of the frame. Typically, the rebar is formed into hoops or half-hoops, and then joined as in the photo above to make a small hoop house. The base may be more rebar, or dimensional wood, or even something more durable like synthetic decking material which looks like wood. One particularly interesting design I saw in person, was a dome shape with a round base. The whole thing was made from rebar which was joined with readily-available U-shaped clamps available at any hardware store. The frame was then covered with a tarp, and wrapped with chicken wire. Each dome had a large hatch, positioned between two of the dome supports and also framed with rebar, which allowed for the human flock owner or caretaker to refill feeders and waterers. The whole thing could be moved by a single person, and it offered wonderful protection against both weather and predators. In that instance, the tractor was used to raise broiler chicks, so no roost materials or nest boxes were needed. That design would just as easily be used for a small layer flock.

The main advantages then would be the lightweight yet strong frame, the excellent mobility, the good protection against predators, and the easy assembly. There are a few disadvantages. If the shelter is used as a chicken tractor, and has no bottom, it's relatively easy for a predator to come along and learn to flip the shelter over, which then gives easy access to the contents. If a tarp is mounted somewhere on the structure, strong winds can do the same thing. So a shelter would need to be anchored against both. A few straps over the top are generally sufficient. The other disadvantage is that the all-metal frame can be an attractant for lightning.

If rebar is used in lightning country, particularly if these structures are going to be out in the middle of an open field or pasture, that can be a consideration. And finally, most chicken tractors made from rebar are assembled with a variety of mechanical fasteners such as bolts, clamps and the like. That means lots of nuts that can work loose over time. This issue can be eliminated right off the batt by using lock nuts which are much less likely to loosen over time.

Another disadvantage would normally be rodent access, given the open bottom. However, this isn't as big an issue with chicken tractors. They are moving so frequently, and often only seasonally, that rodents don't have a chance to tunnel underneath. Furthermore, they can't chew through the metal frame. This is where the coop or chicken tractor's purpose will help determine which risks it will, or will not face, and which issues aren't going to be a problem. Rodents are typically only a mild to moderate issue for field use. If this chicken tractor or coop will be stationary for any length of time, then yes rodents will be a problem and the design should reflect that.

## Chain Link Dog Kennels

*Chickens and roses at <u>Rancho Agua Quieta</u> in Kaufman, TX. A wonderful example of how poultry yards can be a beautiful addition on the landscape. Photo taken by Bryan and Caroline Buck, and is used with permission.*

Chain link dog kennels are another very common material used for chicken coops. Interestingly, these pre-made panels are used in two different ways – as the walls of the coop itself, and/or as the exercise yard next to or around a standalone coop. Let's look at both.

The very attractive above photo shows the birds of <u>Rancho Agua Quieta</u> in Kaufman, Texas, owned by Bryan and Caroline Buck. The chain link enclosure is actually the birds' exercise area, and part of the larger landscaped yard. They actually do have a small coop hidden behind the rose bushes on the left of the photo. Yet the yard itself is big enough for at least one waterer and feeder, which are visible in the background. There is also a large covered area to the back of the yard for rain events. If desired, this yard could be set up with a few roosts and nest boxes, the provision of which would eliminate the need for a coop (at least in mild climate areas). Those areas with heavy rain or snow would need more substantial shelter, but the chain link panels are rigid enough to work well with that arrangement too. Not every poultry yard is lucky enough to be framed by rose bushes. This yard is tidy enough and attractive enough to be considered aesthetically pleasing in almost any neighborhood. Not only is this an example of a nicely set up poultry yard, it shows that anyone with concerns about aesthetics can design their poultry yard to be quite pleasing to the eye.

This approach offers a few more advantages. Chain link panels are generally joined together by clips which are included in the kennel kit, but additional clips can be purchased separately. This means that two or even three kennels can be joined together for much larger areas, as is apparent in the photo above. Additionally, those clips mean that the kennel panels can be taken apart, and put back together, very quickly with just a wrench or pair of pliers. While the panels themselves can be rather unwieldy to move, I have moved them successfully by myself when necessary. And because they're rigid, they are freestanding (to a point), and can serve as walls or attachment points for things like roosts, nest boxes, etc.

The three biggest disadvantages for chain link kennels are their fixed dimensions, their cost, and the fact that rodents can easily burrow underneath them and thus gain access to the coop and/or yard. Let's look at the fixed size issue first. Most chain link kennels come in one of only a few sizes: 4' wide by 10' long, and 6' wide by 10' or 12' long. Furthermore, some kennels are only 4' tall, which would be fine for birds but difficult for people to walk in. These sizes give a decent amount of room for a coop, or a small exercise yard for a small flock. But a large flock will very quickly denude any greenery within the enclosure and turn it into a mudpit during the rainy season. This design then is really best used for flocks up to about two dozen birds each, where predator pressure is quite high. They are almost ideal for urban settings where there's not much space, with lots of cats and dogs nearby, and with neighbors who insist that the pen doesn't become an eyesore. A larger area would be more affordably enclosed with different kinds of fencing.

Speaking of cost, while kennels are easily available at larger hardware stores and most home improvement centers, they are not cheap. They are much more expensive in kit form, than the raw materials would be by themselves. Folks who have experience working with chain link, or who wouldn't mind getting some experience, would be money ahead to buy the raw materials and making their own custom-sized enclosure to their own size specifications. We have done that here and generally been satisfied with the results. An alternative is to keep an eye on the local classified ads and bargain newspapers or websites, such as Craigslist.com, for folks who are selling used chain link kennels. We got our first chain link kennel along with our first dog; the dog's previous owner didn't want either anymore. The second chain link kennel was purchased for a grand total of $50 thanks to a classified ad. Cheap is always better than full price, as long as the materials are still in good shape.

Finally, rodents can and do burrow very easily underneath chain link kennels. Once those burrows are made, they become little rodent highways for raiding feeders. One very easy fix for that, as we discussed in the Rodents section, is to locate the exercise yard and/or coop on sand. Alternately, install a sand "apron" or moat around the boundary of the yard, at least 3' wide. That is sufficient to keep rodents from burrowing underneath.

## Geodesic Domes

*A 10' diameter field pen made from chicken wire or netting stretched over PVC poles. This pen is available in kit form from www.ziptiedomes.com. Photo taken by staff of www.ziptiedomes.com and is used with permission.*

The geodesic dome structure is a relatively recent invention, dating back less than 100 years. The history of the geodesic dome itself is interesting, but beyond the scope of this manual. For folks interested in that history, check out http://en.wikipedia.org/wiki/Geodesic_dome for a good discussion. In the meantime, allow me to summarize the geodesic dome's many structural advantages:

- It is very strong, with a minimum of materials

- It encloses the greatest possible volume per given surface area

- It can be constructed and maintained with an average skill set

- It can be made from a variety of materials, all of which are easily available

- Its round shape makes it very stable, and least affected by strong winds

These structures have a lot of fans, and for good reason. Any given coop needs to be strong, affordable, able to resist weather, and provide ample interior space. So why aren't they used more often?

The heart and soul of the geodesic dome is the joint where the individual structural pieces intersect. That joint, also known as a hub or node, is complex structurally. Not only do multiple pieces come together and need to be fastened, they also do not lay flat. They meet at a slight point, which creates part of the contour of the overall shape. Look at any geodesic dome and subtle points will be visible all over the surface, where these hubs rise slightly higher than the surfaces around them. So thus hubs must be assembled carefully, and exactly..

The additional issue is that the rods used all over the structure, are not all the same length. They are two or three slightly different lengths based on which part of the structure they form. Those lengths are calculated via a moderately complex formula. They need to be exact, and consistent. A lot of variability between rod lengths simply won't work, and the structure can't merely be "nudged" to accommodate a rod too long or too short. So whatever advantages these structures offer, their construction is an exacting business.

Fortunately, interested flock owners and geodesic builders have a number of "helpers" available. First of all, a number of websites exist to provide rod lengths needed for any given radius desired. One such website is here: http://www.desertdomes.com/dome3calc.html. So for instance, if a flock owner wanted to build a geodesic dome that was 20' across, he or she could go to that website and enter a radius of 10', press Submit, and have the calculator spit out the lengths needed for the three different rod types.

Another helper is a series of fasteners, so that the hubs can be formed exactly, reliably, and consistently. So many people are working on ways to simplify this type of structure, that multiple variations on these joints are now commercially available. A list of these brackets is available here: http://www.domerama.com/dome-basics/geodesic-dome-hub-connectors/. Using the calculator listed above, and a set of the available hub connectors, anyone with some DIY experience and interest, can competently build a geodesic dome.

Now that it's a lot easier to build a geodesic dome, what are the advantages and disadvantages of using this approach? As we described above, geodesic domes are extremely lightweight yet strong, and very cost effective in terms of materials used vs. space enclosed. They are extremely resistant to winds, and any covering over the shelter would shed even large amounts of ice, rain and snow with ease. The interior is airy and spacious, plus the full coverage amounts more to an aviary where the birds can safely fly, rather than an exercise yard where they can only run around.

Given all those advantages, are there any disadvantages? Of course. Every approach has its drawbacks and this one has its share. At the top of the list is the shape of the enclosure. Just as that round shape sheds ice and snow so easily, it is very difficult to efficiently cover with any material such as tarps, greenhouse plastic or even chicken wire or bird netting. The pseudo-round shape makes for either a lot of waste of flat materials, or a lot of time spent custom-fitting that covering. The single easiest solution is to use the cheapest material available, and figure to have some waste. Or, in the case of bird netting, simply let it overlap. If using tarps, try using several small tarps instead of one large one, and overlap them so they stay snugged to the structure better.

This structure has similar disadvantages as other enclosure types. If the structure is not on sand, rodents can burrow underneath and get inside. If a flimsy material like chicken wire is used to cover the frame, predators can push, wiggle or chew through. And if the flock density inside the enclosure is too high, the birds can turn that beautiful area into a mud pit unless the whole thing is covered with waterproof material.

One interesting alternative for folks who want a geodesic dome, is the option to simply buy a kit. The photo above is a 10' Zip Tie Dome, available in kit form online at www.ziptiedomes.com. It is a relatively inexpensive kit. I've priced a lot of coop kits over the years and I've built a lot of coops from scratch. I'm not sure I've ever seen a kit offered for such a low cost on a per square foot basis (or cubic foot basis for that matter). For that matter, I'm also not sure I've ever built a coop from scratch for such a low cost. Being made of PVC, these shelters would be ideal for use as seasonal chicken tractors. They're sturdy enough to resist a great deal of bad weather, and they are light enough to be easily moved by one person. This particular shelter only has the one tarp on the top for warm-weather protection from sun and rain, and one cinder block holding it in place. More of the dome could be covered, but more anchoring would then be needed. In the grand scheme of things, that's cheap insurance considering the amount of interior space enclosed by a very rugged structure.

## Converted Barn Stall

*My neighbor very much wanted birds, but a stand-alone coop just didn't suit her property layout. So she converted one of her horse stalls into a chicken stall, and gave them an outside run which can be moved periodically. The chair in the stall is so that she can visit with the birds and unwind after a long day's work away from the property. Photos by the author.*

This may seem an obvious choice, but sometimes it's so obvious we forget it is an option. Wannabe flock owners who have access to standard livestock barns, can often convert some of that space over to poultry use with relatively minor effort. In the above pair of photos, a standard 12'x12' box stall for horses, was very nicely converted for chicken use. In the photo on the left, the flock owner installed a set of storage cubicles in one corner, and fastened fascia boards across the front of each row to serve as landing perches and hold in nesting material. The feeder and waterer are visible in the lower right. The back corner is a very simple roost, made from branches she found on her property, nailed to the two uprights which in turn are secured to the wall. The dirt floor is bone dry and a little bit of straw will keep it dry after the birds move in. She'll be able to sweep up the straw periodically and put it on her garden. Barely visible against the lower far wall, a row of cinder blocks will help keep rodents and predators from burrowing in from the outside. She has decided against a sand moat for the time being, but may use that in the future if the rodents and/or predators get too aggressive. The chair visible in the photo is for her. She not only wanted the birds to provide eggs, but also she simply enjoys them. She found an inexpensive yet comfortable chair, easily cleaned, which will allow her to spend time simply being with the birds and enjoying their company.

The photo on the right shows outside the stall. The chain link yard encloses a nice grassy area for the birds to get some exercise and sunshine each day. A small doggie igloo was included in the yard in case the birds want to have a little hidey spot while outside. And she added two more branches across the yard's two outside corners, to serve as additional perches. It's hard to see in the photo, but the open walls of the stall have been closed off with wire such that the birds can't fly out and predators can't climb in. Also I should point out why she chose this stall, rather than the other two along the back of the building. This stall's two outside walls give her some interesting yard layout options. The flock owner can move this chain link yard around the corner, and give the flock a completely new yard. She can also leave the yard's outer gate open, and let the birds go into either the small yard in the foreground of the photo, or (if the yard's orientation is changed) the large fenced pasture beyond. .A very flexible layout and an example of how to rotate the bird yard amongst different areas, with a minimum of additional equipment.

Not only is this setup extremely inexpensive, it also makes the most of what the flock owner already had. She didn't have to buy the stall separately since it was already on her property, and merely serving as storage. She had originally thought to use the chain link pen by itself as a coop, but she wasn't happy with the amount of space and the mud issues which would have resulted. This hybrid system will provide a nice home for the small flock she wants, with minimal issues.

# SECTION IV: SUPPLEMENTAL INFORMATION

*Three Hamburg hens from the author's egg laying flock, out in a side yard hunting for bugs.*

## *A Few Last Comments*

The backyard flock is experiencing an unexpected, and perhaps unprecedented, revival in 21st Century America. In this age of vacuum-sealed meats, powdered eggs and Styrofoam egg cartons, somehow a growing number of individuals and families want the real thing – a small flock of live birds, producing fresh eggs, right in the backyard. While that might seem a throwback to a bygone era, modern tools are increasingly available to help these new and aspiring flock owners. The number of websites, books and nonprofit groups dedicated to help the small flock owner is large and increasing every day. This manual is a small drop in the bucket of that total body of knowledge.

When I first started writing this manual, I merely wanted to share what we've learned in the last 10 years of small-scale flock ownership. We'd run the whole gamut from starting as pure poultry newbies, looking eagerly at every scrap of information we could find, to being proud owners of a small commercial flock of layers, with egg sales at the farmer's market every week. We'd built, repaired, rebuilt, replaced and redesigned our own coops repeatedly. We'd put up fencing to keep out predators, then put up more fencing when the predators got through what we had. We'd bought in day old chicks and raised them under heat lamps. We'd allowed broody hens to stay on a clutch of eggs, and raise them herself. And here we are, 10 years later, with a fairly diversified amount of information which we'd like to share with those who are also just starting out, or looking for new ideas.

Yet while writing this manual, I became newly aware of how much new information was really out there. It's sobering, and exhilarating, to be amongst those who can help new flock owners get a strong start. So in this section, I'd like to share some of those resources. May these resources be another tool in the new flock owner's toolbox.

## *Additional Resources*

### *Books*

An incredible number of books have come out on the topic of small-scale poultry ownership. Some of those are here, along with a few classics that have been around, and been respected, for a long time.

Storey's Guide To Raising Chickens, by Gail Damerow
This is a very good overall introduction to keeping small-scale flocks, either for hobby or small commercial purposes. Damerow covers all the

basics: breeds, poultry behavior, shelter, health, flock management, egg laying, hatching and chick rearing, poultry exhibition and showing, and meat flock management. If the prospective flock owner does not yet have a good general poultry reference book, this would be an excellent candidate.

Raising Poultry the Modern Way, by Leonard Mercia
This is another excellent introductory text for keeping small-scale flocks. This book is similar to the above except that it covers not only chickens but also turkeys and waterfowl. If a flock owner would like to have more than just chickens, but still needs a good general introductory manual, this would be an excellent choice. I should point out that while I have and still reference this 3rd edition from this author, some of the information has become a bit dated since its release in 1990. A 4th edition of this title has been released by the publishers, however it was written by a different author.

The Small Scale Poultry Flock: An All Natural Approach to Raising Chickens, by Harvey Ussery
This is a very new addition to the flock management library of books. I have not yet read it, but it is recommended by a lot of people whose opinions I respect. Like the first two books above, this text covers a wide range of topics relevant to small flock ownership and management. Unlike the above texts, this very new book covers relatively recent topics such as using electronet and other more modern types of fencing. It also delves quite deeply into the topic of either growing poultry feed or sourcing it cost effectively, which has become a very large concern in recent years. I'll be buying this one for my own collection.

Pastured Poultry Profits, by Joel Salatin
This book is an excellent introduction to the business end of pasture-based poultry operations. Salatin covers both layer flocks and broiler flocks. His book describes his chicken tractor design and usage in great depth, including construction plans. His book also goes into great depth talking about the benefits of having birds on pasture, either as part of a larger rotational management program with mixed livestock, or as a standalone operation. His book is definitely geared towards small-scale commercial production rather than hobbyist flock ownership. However, if a hobbyist is

toying with the idea of getting into egg sales or broiler production, this book would be a very good introduction for doing so.

**Chicken Tractor: The Permaculture Guide to Happy Hens and Healthy Soil**, By Andy Lee and Patricia Foreman
This book is a very good companion to the above text. Like Salatin's book, this focuses on pastured poultry production for small scale commercial operations. However, this book looks at a few different types of tractors and how they can be used for pastured production. It also covers how stationary coops can be used in conjunction with pastured operations. It focuses more on the flock husbandry aspect, rather than the business aspect of pastured poultry. As such, this book would be a better match for the hobbyist flock owner who isn't interested in sales, yet wants to learn more about proven ways to use chicken tractors on the landscape.

**Backyard Chickens' Guide To Coops and Tractors: Planning, Building and Real-Life Advice**, by members of [Backyardchickens.com](Backyardchickens.com)
While I have not read this book per se, I have spent a lot of time on the [backyardchickens.com](backyardchickens.com) website. I have come to appreciate the vast amounts of information available on that website. Of all the websites I visited as part of creating this book, backyardchickens.com was far and away the single best source of pure creative inspiration. Hundreds, if not thousands of people have shared photos and stories about how they have brought small flocks into their lives. As a result, I would expect this book to provide a wealth of information about small household flocks.

## *Websites*

The number of websites about poultry in general, is astounding. Doing a Google search just on "chicken coops" will also yield a lot of results. It can seem overwhelming at first, yet several websites consistently rise to the surface. Here are the best I've found so far:

[www.backyardchickens.com](www.backyardchickens.com)
This website is unique in the sense that it's a big gathering of folks who have taken photos of their coops, to show other people. The website also has a forum where chicken fans can discuss different aspects of flock ownership. I found quite a few examples here of chicken coop plans and

designs, available for a small cost. If nothing else, it's a smorgasbord of coop designs, more than any other single website I've ever found.

www.mypetchicken.com
This is a commercial site, offering a wide variety of products for sale for the hobbyist flock owner. Unlike many other commercial sites for poultry supplies, this one assumes that the flock owner has birds for enthusiast purposes, as pets or yard ornaments moreso than as pay-their-way livestock. Many flock owners have birds for purely hobbyist reasons, and this site caters to their needs a little better than other websites which are geared more for the commercial flock owner. I suggest that small-scale flock owners check this site, but also check commercial sites as well, if only to shop for the best prices on any particular item.

www.urbanchickens.org
This website is dedicated to the legalization of keeping small flocks in urban areas. The website provides up-to-date information on legalization efforts around the country, along with information for urban owners or those working for urban legalization.

www.urbanchickens.net
Not to be confused with the similar-named website above, this particular website is actually a weblog. The blog appears to be written by one individual, and concentrates on topics relevant to urban chicken ownership. The blog page includes ads related to same.

http://madcitychickens.com
This website was created by a group of individuals whose stated goal is to help folks legalize small flock ownership in various urban areas around the USA. The site has information on the methods for changing laws, along with information about those efforts which have already been successful (along with what was learned in the process). These folks are often listed on other websites as THE go-to source of information about how to successfully change municipal code to allow for chicken ownership.

http://citychickens.com/

This is a website dedicated to urban small flock ownership. It offers information on a variety of topics, including breeds, egg-laying, coops, books, etc.

## Handouts, Reports and Other Publications

ATTRA is a nonprofit organization devoted to gathering and distributing information about sustainable agriculture. In particular, that fine organization researches and publishes reports on specific topics related to sustainable agricultural production. In our context of poultry coops and poultry production, they have several publications of interest to both new and existing flock owners. Not all of these will be relevant to every backyard flock owner. Some of them are instead intended for small scale commercial producers. However, they are all very readable, and provide a lot of solid, well-researched information for what works, and what doesn't, in both meat and egg flocks. And best of all, the reports are free.

### Pastured Layers
https://attra.ncat.org/uofa/docs/PasturedLayers.pptx
This ATTRA PowerPoint presentation covers a variety of egg layer breeds, and various aspects of egg layer flock management from chick raising to egg marketing. The presentation also provides some information on how to evaluate eggshell defects, how to clean eggs and how to store them safely.

### Range Poultry Housing
https://attra.ncat.org/attra-pub/summaries/summary.php?pub=233
This article covers small-scale commercial pastured poultry housing. As such, it is a very good review for small flock owners who are either ready to scale up, or dream about doing so someday. A few of our coop plans would fall within small scale commercial flock housing, but this article goes into much more depth on various designs for field production.

### Poultry Equipment for Alternative Production
https://attra.ncat.org/attra-pub/summaries/summary.php?pub=230
This is an ATTRA report which can either be read online in HTML format, or downloaded as a PDF. The article describes the equipment, housing, fencing, nest boxes and other equipment needed for pastured

layer production. While this report may be intended for small scale commercial production, it will give even an urban home flock owner some good ideas on how to improve efficiency, and boost egg production.

Urban Poultry
https://attra.ncat.org/attra-pub/poultry/urban.html
This one-page article is intended for city residents who are just starting to consider flock ownership. It provides a few guidelines complying with local regulations, covers the basics of what birds need to stay healthy, and describes what a flock owner can expect from his/her flock owning experience.

## *Conclusion*

In closing, I sincerely hope that the information, links and images in this manual, all come together to give prospective flock owners a good basis upon which to design their own coop. So very many people want to get started with chicken flock ownership, and they are overwhelmed with all the options for how to house them. Hopefully this guide has given some clarity to that situation. Here's hoping that everyone who aspires to owning a flock, can use this manual to go out and provide the best housing for their particular wants and needs.

I would invite readers to let me know their comments, questions, suggestions and/or criticisms of this manual. Even as I was finishing up the book, I became aware of so many other options which could have been included, but weren't only because the book was too far along to include them at the last minute. So there's a decent chance that I will someday release a Second Edition. That would be the perfect time to make improvements, corrections, clarifications, etc. Reader comments can only help to make any future edition better.

Kathryn Kerby
www.frogchorusfarm.com
January 16, 2014

## ABOUT THE AUTHOR

Ms. Kerby spent much of her teenage years on horseback, haunting the old ranchlands in then-rural Douglas County, Colorado. After high school she attended the University of Colorado-Boulder where she earned her Bachelor's of Science in Environmental Biology, with an additional focus on technical writing and scientific journalism. Between 1988 and 2011, she held a variety of jobs where either her scientific and/or journalistic skills were exercised regularly. During that time she also lived in six different states, from one coast to the other and several points between. In 2000 she and her husband purchased an old rundown property and dilapidated farmhouse in western Washington state, and started the hard work of restoring the property to full production. That work is still underway. They added their first poultry flock in 2002 and have had birds ever since. Farming and farm-related products became her sole source of income in 2011. As of this writing in 2014, the farm has become a very well diversified livestock operation, with hay, field crops, forestry and garden also competing for attention. Visit the farm's website at www.frogchorusfarm.com. Her favorite things besides farming, gardening and working with her animals, are drinking a hot cup of tea, reading an old book, listening to various types of music, or watching either westerns or sci-fi movies.

## PHOTO AND ILLUSTRATION CREDITS

All images and photographs in this book were used with prior written permission from the photographer and/or owner.

**Book Location:** Photo ID, Photographer or Artist, Property Owner and Location

**Front Cover,** *clockwise from top left:*
Standalone Pallets Coop, Butch Bridges, Lone Grove, OK
Cattle Panel Chicken Tractors, Kinikin Natural Foods, Gunnison, CO
Coop Construction Plans, Louisiana State Univ / AgCenter, www.lsuagcenter.com
Geodesic Dome Field Pen, www.zipticdomes.com
Hoop House Chicken Coop, Adrienne Cahoon photographer, Kent Solberg owner, Seven Pines Farm & Fence, Verndale, MN
Chain Link Chicken Yard, Bryan and Caroline Buck, Rancho Agua Quieta, Kaufman, TX

**Introduction:** Mixed Breed Hen, Kathryn Kerby, Frog Chorus Farm, Snohomish, WA
**Section I:** Hen and brood, Kathryn Kerby, Frog Chorus Farm, Snohomish, WA
**Section II:** Hen on roost, Kathryn Kerby, Frog Chorus Farm, Snohomish, WA
**Section II:** Plan #1, Louisiana State Univ / AgCenter, www.lsuagcenter.com
**Section II:** Plan #2, Louisiana State Univ / AgCenter, www.lsuagcenter.com
**Section II:** Plan #3, Louisiana State Univ / AgCenter, www.lsuagcenter.com
**Section II:** Plan #4, Louisiana State Univ / AgCenter, www.lsuagcenter.com
**Section II:** Plan #5, Louisiana State Univ / AgCenter, www.lsuagcenter.com
**Section II:** Plan #6, Louisiana State Univ / AgCenter, www.lsuagcenter.com
**Section II:** Plan #7, Louisiana State Univ / AgCenter, www.lsuagcenter.com
**Section II:** Plan #8, Louisiana State Univ / AgCenter, www.lsuagcenter.com
**Section III:** Winter Hens, Kathryn Kerby, Frog Chorus Farm, Snohomish, WA
**Section III:** Car Canopy, Kathryn Kerby, Frog Chorus Farm, Snohomish, WA
**Section III:** Greenhouse, Adrienne Cahoon, Kent Solberg/Seven Pines Farm & Fence, Verndale, MN
**Section III:** Truck Bed With Canopy, Kathryn Kerby, Frog Chorus Farm, Snohomish, WA
**Section III:** Shipping Crates, Olde Tyme Coops, Olde Tyme Coops
**Section III:** Livestock Trailers, Debbie Dern, Dern Residence, Snohomish, WA
**Section III:** Cattle Panels and Tarps, Kinikin Natural Foods, Kinikin Natural Foods, Gunnison, CO
**Section III:** Metal Erector Frames, www.shelters-to-go.com, www.shelters-to-go.com
**Section III:** Pallets, Butch Bridges, Bridges Residence, Lone Grove, OK
**Section III:** Rebar Frame, Walter Jeffries, Sugar Mountain Farm, West Topsham, VT
**Section III:** Chain Link Dog Kennels, Bryan and Caroline Buck, Rancho Agua Quieta, Kaufman, TX
**Section III:** Geodesic Domes, www.ziptiedomes.com, www.ziptiedomes.com
**Section III:** Converted Barn Stall, Debbie Dern, Dern Residence, Snohomish, WA
**Section IV:** Three Foraging Hens, Kathryn Kerby, Frog Chorus Farm, Snohomish, WA